The Restless Universe

THE RESTLESS UNIVERSE

Understanding X-ray Astronomy in the
Age of Chandra *and* Newton

Eric M. Schlegel

OXFORD
UNIVERSITY PRESS

2002

OXFORD

UNIVERSITY PRESS

Oxford   New York
Auckland   Bangkok   Buenos Aires   Cape Town   Chennai
Dar es Salaam   Delhi   Hong Kong   Istanbul   Karachi   Kolkata
Kuala Lumpur   Madrid   Melbourne   Mexico City   Mumbai   Nairobi
São Paulo   Shanghai   Singapore   Taipei   Tokyo   Toronto

and an associated company in Berlin

Published by Oxford University Press, Inc.
198 Madison Avenue, New York, New York 10016

www.oup.com

Library of Congress Cataloging-in-Publication Data

Schlegel, Eric M. (Eric Matthew), 1954–.
The restless universe : understanding X-ray astronomy
in the age of Chandra and Newton / Eric M. Schlegel.
p.   cm.
Includes bibliographical references and index.
ISBN 0-19-514847-9 (cloth)
1. X-ray astronomy. I. Title.
QB472 .S35   2002   522´.6863—dc21
2002072755

ISBN 0-19-514847-9

1 3 5 7 9 8 6 4 2

Printed in the United States of America
on acid-free paper

*To the memory of my father, William H. Schlegel (1909–2000),
for much-needed advice at an unexpected time;
my mother, Jane S. Schlegel, for continually asking,
"When will I see your name in the paper?" to which I answer,
"Will this book do instead?"; and my wife, Lisa M. Schlegel,
for all her support and encouragement.*

CONTENTS

Preface IX

Overview XIII

1. "By three methods we may learn wisdom . . ."   3
   (in which the Reader meets three X-ray satellites)

2. ". . . and know the place for the first time."   12
   (in which the Reader learns how to locate celestial sources of X rays)

3. ". . . every solution serves only to sharpen the problem, to show us
   more clearly what we are up against . . ."   37
   (in which the Reader sharpens those locations)

4. "Sometimes you get shown the light in the strangest of places
   if you look at it right."   59
   (in which the Reader learns about brightness and luminosity)

5. *Veritatem dies apertit.* (Time discovers the truth.)   76
   (in which the Reader sees the importance of arrival times)

6. ". . . a spectrum is worth a thousand pictures."   92
   (in which the Reader encounters spectroscopy)

7. "We're all nothing but unified arrangements of atoms . . ."   114
   (in which the Reader learns how spectroscopy connects to atoms)

8. "If you have an important point to make, don't try to be subtle
   or clever. Use a pile driver . . ."   129
   (in which the Reader receives a summary of X-ray astronomy)

9. "Destiny is no matter of chance. It is a matter of choice . . ."   143
   (in which the Reader learns about costs and choices)

10. "I like the dreams of the future better than the history of the past."   154
    (in which the Reader gets a glimpse of future satellites)

Notes 165

Bibliography 191

Suggested Readings 197

Index 199

Preface

Scientists exploring the field of X-ray astronomy are in the midst of a time that will not come again; we enjoy the excitement of what is often called an age of discovery.[1] Dr. Carl Sagan said that there is only one generation that gets to see things for the first time (in his case, the surfaces of the planets). This is a unique time because prior generations knew little about X rays, and subsequent generations will view today's amazing discoveries as history and as stepping-stones for yet greater discoveries.

Science advances in phases, starting from pure discovery, or "un-covery," and ending with a mature field in which most of the questions have been answered and little additional progress is possible. These advances do not parade steadily forward in time. Consider, instead, the approach fans of jigsaw puzzles follow. Assemblers first locate straight-edged pieces to build the frame. With the frame in place, they see the range of colors and note potentially easy areas on which to focus attention. They sift through the box, searching for those pieces first. When the easy areas are done, the task shifts to filling in details, all the blue sky pieces, for example.

In 1960, a relatively slim astronomy book contained essentially all of our knowledge of the planets in our solar system. By 1990, however, the discovery phase of planetary astronomy described by Sagan had essentially ended. We peered at the surface of Mars from three landers (Vikings 1 and 2 and Pathfinder), gazed upon the unique surfaces of the Galilean satellites (Io, Europa, Ganymede, and Callisto) of Jupiter, imaged the rings of Saturn and the outer planets Uranus and Neptune with their moons, rings, and atmospheric spots, discovered a moon of Pluto, and flew a satellite through the tail of a comet. Images of all the planets save one (Pluto) existed and could be bought in poster shops. Complete books for each planet had been written, summarizing the missions of the 1970s and 1980s. Arguably, those who paid attention to the planetary missions from the late 1960s to the late 1980s or early 1990s form the generation to which Sagan referred.

In X-ray astronomy, the period of our "first look" started with the Einstein Observatory (1979-81) and will end sometime in the next 10 to 20 years. Einstein provided the very first images and spectra of the X-ray universe, but for a small number of objects. By the time the Chandra and Newton missions end, we will have a robust inventory. Future generations of X-ray observatories will likely be designed for specific experiments or observational goals instead of functioning as generic observatories.

This book sends the reader on a journey, one that encompasses the entire universe. Instead of a path that leads from Earth outward, this book explores a different route by describing our view of the universe if we study the X rays that arrive from astrophysical objects.

Several approaches exist for a book such as this one. I could present the results of the discoveries of the past 30 to 40 years. I could take readers through the history of X-ray astronomy or focus on how the increasingly sophisticated instruments have allowed detailed explorations of the X-ray universe. I chose to present threads from each of these approaches; I hope I have woven a decent cloth. I do not present everything that has occurred in X-ray astronomy during the past three decades, because the resulting book would be used only as a doorstop. I also do not present each and every "gee whiz" discovery, because many would be obsolete by the time this book appears in print. Instead, I aim to provide a foundation for further learning. The threads include the history of X-ray astronomy; the hardware used to detect X rays; the satellites, past, present, and future, that have been flown to collect the data; how we interpret the data; and most particularly, the science we have learned, as well as speculations about what we will learn. I have also not attempted to place every up-to-the-minute result here, particularly since some of the most interesting science will be a complete surprise.[2]

I have benefited from countless conversations, about X-ray astronomy and its discoveries, with colleagues at science meetings and at the two places I've worked during the past ten years (the NASA-Goddard Space Flight Center and the Smithsonian Astrophysical Observatory). These colleagues are too numerous for me to identify individually; I thank all for lively discussions, whether they remember them or not. I also gained knowledge from many people connected with the Chandra project. A project the size of Chandra requires many hundreds of people, from administrative assistants to scientists, engineers, and project managers. Space constraints preclude listing even a small fraction of these people. The list of institutions significantly involved in Chandra's design and construction is itself long: NASA's Marshall Space Flight Center (Huntsville, Alabama); the Office of Space Science at NASA headquarters (Washington, D.C.); the TRW Space and Electronic Group (Redondo Beach, California); Raytheon Optical Systems, Inc. (Danbury, Connecticut), now a division of Goodrich Corporation; Optical Coating Laboratory, Inc. (Santa Rosa, California); Eastman Kodak Company (Rochester, New York); Pennsylvania State University (University Park, Pennsylvania); Space Research Organization Netherlands (Utrecht, Netherlands); Max Planck Institute (Garching, Germany); Massachusetts Institute of Technology and Smithsonian

Astrophysical Observatory (Cambridge, Massachusetts), and Ball Aerospace and Technologies Corporation (Boulder, Colorado).

I thank my agent, Jeanne Hanson, and my editor, Kirk Jensen, for seeing the potential in an early draft of this book. Thanks to the copyeditor, Jane Taylor, for catching several recurrently missed mistakes, and to the production editor, Joellyn Ausanka and the overall compositor, Anne Holmes, for turning a stack of manuscript and illustration pages into a sharp-looking book. I hope the words live up to their efforts; any errors that remain are mine.

I especially thank my wife, Lisa, for her love and encouragement on those days when the universe seemed too amazing to be described by mere words.

Overview

Brevis esse laboro, obscurus fio.
(When I labor to be brief, I become obscure.)
—Horace

X rays are light, light that is fundamentally no different from the optical light that enters our eyes or the radio light that carries the signal from our favorite radio stations to our cars.[1] All of these—X rays, visible light, radio light, and more— may be described as waves, specifically electromagnetic waves,[2] first described by the Scottish physicist James Clerk Maxwell (1831–79) in the 1860s. At that time, X rays were unknown, not to be discovered for another 30 years. As we shall see, X rays give us a picture of the universe very different from the one available to our eyes and our optical telescopes.

Anyone who has been to a dentist or to a doctor to repair a broken arm knows what an X ray is. Unfortunately, the term *X ray*, so applied, refers to the piece of film illuminated by X rays. Mention "X-ray astronomy" to people and too often they think we send beams of X rays *into* space to obtain an "X ray" of an object. Many of us who carry out research on X-ray energies quickly learn to indicate that the X rays we study come *from* somewhere in the universe. That comment is usu- ally accompanied by a motion of the arm that starts with it fully extended and ends with it close to the body.

So what are X rays? That we cannot see them with our eyes is irrelevant. Each of the major areas of scientific knowledge is hip-deep in examples of insight gained by looking at things we cannot see. X rays bathe Earth each second, arriving from everywhere in the universe. Long before anyone discovered them, X rays carried their energetic message, and they will do so long after Earth has ceased to exist. What we have learned in the past 30 years about the X-ray universe is astounding. The space missions that have been and gone, and those that still collect data, have largely defined the straight-edged pieces of the puzzle. The view of the universe given to us by X rays may turn out to be absolutely critical to our understanding.

Not only can our eyes not detect X rays, but the atmosphere of Earth blocks X rays from even reaching the ground. If our atmosphere did not do so, life would likely never have started on this planet. The atmosphere protects us from the cellular damage that the absorption of X rays produces. We need telescopes and detectors, sensitive to X-ray light and launched on satellites above Earth's atmosphere, to study the X-ray universe. The Chandra X-ray Observatory,[3] Newton, and Astro-E, the three satellites discussed in chapter 1, are the latest in a series of observatories to explore the X-ray universe, continuing the endeavor of the past 35 years.

Chandra is the third of NASA's "Great Observatories." The first is the Hubble Space Telescope,[4] the second is the Compton Gamma-ray Observatory.[5] The fourth and final Great Observatory will be the Space Infrared Telescope Facility (SIRTF), currently scheduled for launch in late 2002. Each of these observatories has made, or will make, fundamental contributions to our understanding of our universe. The Compton Observatory for the first time made the discovery and observation of sources emitting gamma rays relatively easy. Data returned from the Hubble Space Telescope clearly excite everyone who sees them. Few can look at the images of the Eagle Nebula, for example, with its tall pillars of dark matter surrounded by glowing, ionized gas, without wonder.

Chandra, as an observatory, is to X rays as the Hubble Space Telescope is to visible light. The results from Chandra will continue to transform our understanding of our universe, if our experiences with Hubble are a guide.

The Restless Universe

1

"By three methods we may learn wisdom . . ."

—Confucius

Chandra, Newton, and Astro-E: three satellites, all dedicated to the study of X rays. Two and a half billion dollars spent. What is so important about X rays from space? Why were three satellites built by three different space agencies representing different countries? Astronomers designed and built only one satellite to study the visible universe (the Hubble Space Telescope). What makes the X-ray universe different? What are the differences among the three satellites? What have we learned from previous X-ray missions? Why do X-ray images look so different from those returned by the Hubble Space Telescope? Why was the loss of Astro-E so devastating to X-ray astrophysics?[1] Will Chandra, Newton, or Astro-E answer all of our questions about the X-ray universe?[2] What are X rays?

Launch: The Chandra X-ray Observatory

On 20 July 1999,[3] at 12:30 A.M. eastern daylight time (EDT), the first launch attempt of the space shuttle Columbia, carrying the Chandra X-ray Observatory, halted at T–7[4] seconds. A hydrogen sensor in the hazardous-gas-detection system surged to 640 parts per million (ppm) from a normal level of about 110 ppm. Engineers had designed the sensor to sample the air in the aft engine compartment every eight seconds. At T–16 seconds, the sensor had reported its first high reading. An engineer on the launch team waited for the next sample at T–8 seconds. When it reported the second high reading, he manually executed a launch abort, stopping the countdown about three seconds before main-engine ignition. Subsequent examination of the data revealed a faulty sensor. The backup sensor, although less sensitive than the primary one, never showed any increase in the gas level.

Forty-eight hours later, on July 22, after the investigation of the gas detection sensors, after the replacement of the ignitors,[5] and after the refilling of the propellants, the shuttle again stood bathed in spotlights. The second launch countdown proceeded smoothly to the planned hold at T–20 minutes. All launches have built-in holds, usually lasting 10 minutes, that provide the team sufficient time to review all the sensors and checklists to be certain that the shuttle, rockets, and crew are ready to go. Toward the end of each hold, the launch director polls the team leaders for a final "go" recommendation to lift the hold.

On July 22, the weather team reported the existence of a thunderstorm eight miles from the launch site. The lightning protocol allows no lightning within 20 miles of the launch complex, or no lightning within 10 miles plus at least 15 minutes of elapsed time since the last detected stroke.

Flight controllers restarted the countdown clock at the end of the planned hold with the understanding that another hold would occur at T–5 minutes until the weather cleared. When the launch window opened at 12:26 A.M., the thunderstorm cell was still present. At 1:11 A.M., the launch director announced that an extra six minutes had been added to the launch window, extending it to 1:24 A.M. Just as the launch team readied to resume the count after the T–5 hold, a lightning strike occurred eight miles from the shuttle. The launch director immediately announced the second launch abort.

On July 23, 1999, at 12:31 A.M. EDT, Columbia lifted off into clear skies (Fig. 1.1). Five seconds after liftoff, one of the electrical buses short-circuited, causing a loss of two of the engine controllers. Each shuttle engine has two separate controllers, so the liftoff proceeded. Had the short occurred before engine ignition, the launch director would have aborted the launch. In addition, had one more engine controller short-circuited during takeoff, shuttle commander Eileen Collins and pilot Jeffrey Ashby would have been the first crew to attempt to abort a takeoff and land the shuttle at the backup landing site in Banjul, Gambia, on the west coast of Africa.

Columbia reached an orbit seven miles lower than had been calculated. The flight team later traced the likely cause of the lower orbit to a fuel leak discovered in the aft engine compartment.[6] The shuttle's crew deployed the Chandra Observatory seven hours after launch (Fig. 1.2). During the subsequent two weeks, ground controllers gradually placed the observatory into its final orbit and unfurled the solar-

Fig. 1.1.   July 23, 1999, 12:31 A.M. eastern daylight time: the launch of the space shuttle Columbia carrying the Chandra X-ray Observatory. Eileen Collins commanded the shuttle and the crew of four astronauts. (Image courtesy of the public image launch archives at the NASA-Kennedy Space Center.)

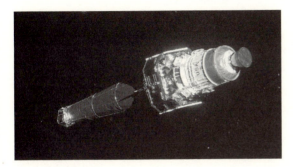

Fig. 1.2. The deployment of Chandra occurred seven hours after launch. Chandra is one of the largest payloads ever carried aloft by the shuttle. The image appears distorted because a portion of Chandra lies in shadow. (Image courtesy of the public image launch archives at the NASA-Kennedy Space Center.)

cell panels (Fig. 1.3). Prior to the observatory becoming operational, all of the parts used to construct Chandra had to first "outgas," or lose the moisture accumulated while in the atmosphere of Earth, to the vacuum of space. Outgassing is necessary before any telescope is opened in space; a failure to outgas can cause the accumulated moisture to freeze onto the telescope mirrors and detectors, considerably reducing their effectiveness. Several weeks of additional work were necessary before confirmation of Chandra's status as an observatory, but it was finally in orbit.

More than twenty years had elapsed from the time NASA received the first proposal to build a high-resolution X-ray telescope. During that time, countless design meetings, months of design reviews, years of work, and four launch slips

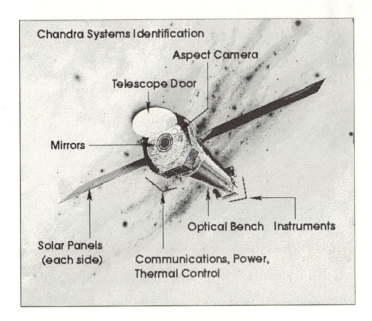

Fig. 1.3. An artist's sketch of Chandra in orbit with its solar panels deployed. The individual parts of the satellite have been labeled. (Artist's concept courtesy of TRW, Inc.)

had occurred. Thousands of people had contributed to placing the observatory into orbit: project and program managers and their administrative assistants; budget teams; mechanical, thermal, and electrical engineers; data aides and technical assistants; members of test teams; and scientists.

Chandra's first observation occurred just after the outer sunshade door, built to protect the sensitive instruments from the overpowering light of the Sun, opened on command on August 12, 1999. The official "First Light" would take place after a check that the mirrors and detectors were working correctly. The prime instrument, ACIS (Advanced Charge-Coupled Device Imaging Spectrometer), imaged a

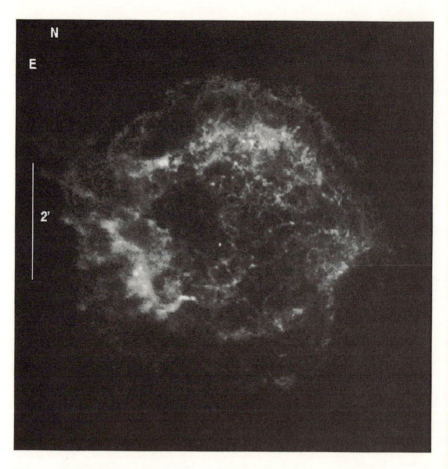

Fig. 1.4. The First Light image of the supernova remnant Cassiopeia A. On the left is the normal view; on the right is a view with black and white inverted. The image is displayed inverted because black on white offers higher contrast of the details; in addition, an excess of black ink sometimes bleeds into the smaller, whiter areas during image production, particularly if image features are thin and narrow. For both reasons, most astronomical images are displayed in the inverted manner, and this book follows that convention. All figures include a scale bar somewhere in the figure, usually on

source lying just off the optical axis of the telescope. Chandra's value lies in its image quality, known as the point-spread function.[7] The size of the point-spread function, when measured after the data were received in the Operations Center, proved that Chandra focused X rays more sharply than had any previous X-ray telescope. This measurement occurred even before attempts were made to sharpen the focus or correct for any drift of the pointing direction.

The official First Light image, obtained on August 19, 1999, is stunning (Fig. 1.4). Cassiopeia A (Cas A) is a supernova remnant, the grave marker of an exploded star.[8] The light from the explosion reached Earth in the year 1670. The

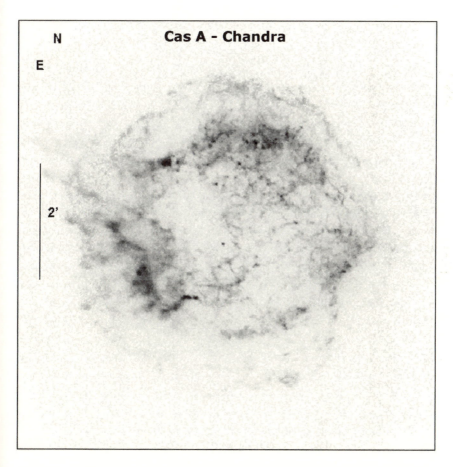

the left, to indicate the size of the object. The scale is usually in minutes of arc (one minute of arc equals 1/60th of a degree) but occasionally is much smaller. In the upper left corner of all figures are the direction designations: the top of the figure is the north edge; east lies to the left. This orients the object as an observer would see it in the sky. The small dot at the center of Cas A is the proposed corpse of the star that blew up to form the remnant. (Image produced from data obtained from the Chandra public data archive at the Smithsonian Astrophysical Observatory.)

distance to Cas A is estimated to be 10,000 light-years, so the star actually exploded about 8000 B.C.; the light traveled for 10,000 years before reaching Earth. The X-ray image shows filaments of hot gas from the exploded star as well as a dot of light at the center of the remnant. That dot is likely the actual carcass of the star, and it had not been detected by any other telescope at any wavelength. So far as we know, essentially two paths exist for a star that is about to explode. The path taken depends on the mass of the star—the quantity of material contained in it. Astrophysicists believe that lower-mass stars explode completely, leaving no stellar corpse behind. For the higher-mass stars, however, the inner layers collapse to form a neutron star or black hole; the collapse blows off the outer layers. Some of the hot filaments in Cas A show evidence of oxygen, silicon, and iron. Combined with the presence of a pointlike source as a candidate for the stellar corpse, Cas A is the gravestone of a high-mass star. Investigations into the point source commenced as soon as the First Light image appeared.

There's more to learn from the First Light image. Examine the figure closely and you will see a faint plateau of light that lies just outside the filaments. This is the expanding shock wave. The explosion of a star not only disrupts the star itself, but also creates a shock wave that expands outward. The expanding shock stirs the gas that lies between the stars. Under the correct conditions, some of that stirred gas will collapse and form new stars. By measuring the speed with which the shock moves—for example, by obtaining a second image several years later and measuring the increased diameter of the shock—we can obtain a better estimate of the age of the remnant. For Cas A, the remnant's age happens to be relatively well known. For other remnants, however, that information could allow us to narrow our search in the historical records for notes indicating the first appearance of the supernova in our sky.[9]

Launch: XMM-Newton

In contrast to the launch of Chandra, that of the European X-ray satellite, the X-ray Multi-Mirror Mission (XMM), from Kourou in French Guiana, South America, on December 10, 1999, proceeded without a hitch (Fig. 1.5). The rocket carrying the satellite aloft was the Ariane 5, a new rocket for the European Space Agency with heavy-lift capability. The launch, while flawless, was not without tension: the test launch of the Ariane 5 had exploded shortly after ignition of the main engines. For the European Space Agency, the XMM launch was a triumph because XMM, although a pure science satellite, represented its first commercial payload.

Controllers on the ground gradually raised XMM's orbit so that, at its closest approach to Earth, it was about 7,000 kilometers high; at the farthest point, it was about 114,000 kilometers from the surface. By December 15, the solar panels were deployed and several of its instruments had been powered up (Fig. 1.6). Mission

Fig. 1.5.   December 10, 1999: the launch of the European equivalent of Chandra: XMM-Newton. This launch took place from the European launch complex in French Guiana and used the heavy-lift capability of the new Ariane 5 rocket. (Image used courtesy of the European Space Agency.)

controllers scheduled full activation of the satellite to occur after the date rollover from 1999 to 2000 just in case any Year 2000 problems were uncovered. By mid-February 2000, all of the instruments had been tested and the official First Light images obtained. XMM was renamed to honor Isaac Newton.

The European Space Agency released the official First Light images to the press on February 9, 2000. The first data were quite impressive. An image (Color Fig. 1 [see insert]) of one of our galaxy's near-est neighbors, the Large Magellanic Cloud, shows the very active region of 30 Doradus, an area known for its abundant births and deaths of stars.[10] The large cir-

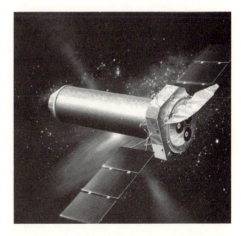

Fig. 1.6.   An artist's sketch of Newton after it reached its final orbit. (Image used courtesy of the European Space Agency.)

cular arc near the center of the image is 30 Doradus C, a supernova remnant. Near the center of the arc is a faint point source. This point source, if it is confirmed to lie at the distance of the Large Magellanic Cloud, is probably the corpse of the star that exploded. The bright object east of 30 Doradus C is the supernova remnant

N157B.[11] This object is similar to the Crab Nebula in our galaxy, the remnant of a star that people observed exploding in A.D. 1054. Another, better known, supernova remnant lives nearby: the remains of the supernova of 1987. SN1987A was the closest supernova in more than 300 years. It is the bright spot southwest of 30 Doradus C. Finally, the faint point sources, particularly those to the southeast of 30 Doradus C, are probably background active galaxies shining through the Large Magellanic Cloud.[12]

Another image (Color Fig. 2 [see insert]) shows a compact group of galaxies known as Hickson Compact Group (HCG) 16, about 170 million light-years away.[13] Seven galaxies make up the group. All the galaxies in the group show evidence of mergers or collisions. Whether galaxies collide is not a contentious issue because we know nearby galaxies show considerable evidence of past collisions. Collisions among galaxies rob both victims of the gas and dust necessary to make new stars, instead spraying the gas and dust away. Mergers replenish the gas and dust as the nuclei of the colliding galaxies gradually sink into a single gravity well. A critical question of current research in astrophysics is the degree of evolution of the structures in the universe. As we look ever deeper, we look ever further back in time. By looking at the earliest galaxies and comparing them with nearby galaxies, astronomers attempt to measure that evolution. The number of mergers and collisions is one measurement.

The Newton image of HCG 16 shows four of the primary galaxies in the group. In the figure, blue designates all X rays with energies above three kilovolts ("hard" X rays; a kilovolt is a unit of energy), while red signifies X rays with energies around 0.8 kilovolts ("soft" X rays). The westernmost galaxy, NGC 833, shows a bluer color, so it has a hard spectrum. Astronomers assign these colors in a consistent but arbitrary manner as a quick method to compare objects in an image such as HCG 16. Note that each of the galaxies shows a "core-halo" structure: the cores appear to be brighter, perhaps indicating gas and dust fueling a central nucleus containing a galactic black hole. The nuclei of galaxies such as those in HCG 16 fall between those of normal galaxies, which have inactive nuclei, and active galaxies, where the nuclei have high luminosities.

Launch: Astro-E

The Japanese Space Agency, late in 1998, scheduled the launch of the Japanese-American mission Astro-E (Fig. 1.7) for February 8, 2000. The Japanese are noted for scheduling a launch date and time a year or more in advance and sticking to the schedule. They have done so on four previous occasions for their X-ray satellites alone. For the Astro-E launch, high winds at the launch site delayed one launch attempt. A second was called off moments before ignition because of a problem with a tracking station.

Fig. 1.7. Sketch of the Japanese-American satellite Astro-E. (Image obtained from the public archives at the High Energy Astrophysics Science Archive Research Center, NASA-Goddard Space Flight Center.)

On February 10, 2000, at 10:30 A.M. Japan standard time, the launch controllers ignited the main rocket engine. At T+25 seconds, sensors in the rocket detected anomalous vibrations. At T+41 seconds, ceramic heat shields in the first-stage

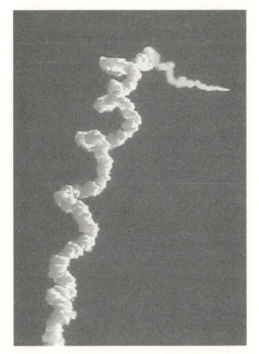

nozzle broke off, damaging the thrust control of the nozzle. Twenty seconds later, the rocket carrying Astro-E veered off course (Fig. 1.8). The second and third stages attempted to compensate, but the thrusts from those stages were too small for the job. Astro-E ended up in an "orbit" that was at most 80 miles high. Earth's atmosphere at that altitude is sufficiently thick that friction between the satellite and air brought the satellite down before it had completed a single orbit. After 6 years of work by thousands of people, Astro-E became a very expensive "shooting star" somewhere over east Africa and Tibet or western China.[14] It was a significant loss for X-ray astrophysics.

Fig. 1.8. Launch of the Japanese-American satellite Astro-E. The corkscrew-shaped exhaust trail is a certain sign of trouble. (Image used by permission of Dr. Damian Audley.)

2

"...and know the place
for the first time."

—T. S. Eliot, *Four Quartets*

X rays teach us about the universe. For the moment, assume that an X ray is a photon emitted by objects in the universe and each X ray carries information about those objects. Set aside for now the nature of photons, X ray and otherwise; instead, assume that the sources of X rays are stars or galaxies or planets. We will have to study these objects to learn which actually emit X-rays.

What information do X rays give us about the source object? We may quantify only four characteristics:[1] the position in the sky from which X rays arrive, their times of arrival, the brightness of the source, and the energies of the X rays. We really only measure position, arrival time, and energy for a particular X ray. By considering all the events collected during some defined interval of time, we determine brightness. Although these characteristics are interdependent, analysis of each provides information, so for now, consider them as separate quantities. The identical quantities may be measured for any photon from radio to gamma ray if the detector is designed appropriately.

Over the next few chapters, we'll look at each quantity in turn. Journalists learn, as one of their first lessons, the six parts of a good opening paragraph for a newspaper article: who, what, where, when, how, and why. Think of science as a search for the same six parts. For astronomers, the four quantities of position, brightness, time, and energy provide measures similar to who, what, where, and when. We use that information to infer the how and the why.

Start with the position of a detected photon. Look closely at the image of Cas A obtained by the X-ray satellite ROSAT (Fig. 2.1).[2] Compare this image with the image from Chandra (Fig. 2.2); the image from Chandra is sharper. Yet the images clearly do not have the smooth appearance of an optical image. Instead, the X-ray images look fuzzy and grainy, like sand sprinkled on black velvet. Why the difference between these images?

One Friday afternoon in November 1895, Wilhelm Roentgen,[3] a German physicist, discovered X rays.[4] Roentgen (Fig. 2.3) called the new, invisible light "X strahlen,"

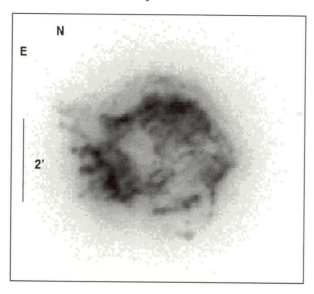

Fig. 2.1. Cas A as seen by the High Resolution Imager on board ROSAT. Compare this image with the next one. The High Resolution Imager was designed to provide the best position and spatial resolution available at the time of its launch in June 1990. (Image generated from data obtained from the ROSAT data archive, NASA-Goddard Space Flight Center.)

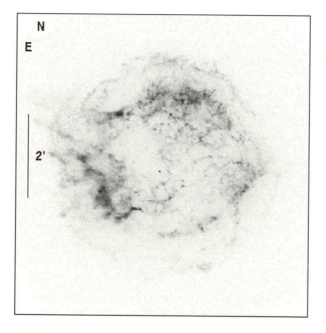

Fig. 2.2. Cas A as seen by Chandra. The ability to locate the precise source of emission of an X ray is about ten times better in this image than in the previous one. (Image generated from data obtained from the Chandra public data archive at the Smithsonian Astrophysical Observatory.)

which, in English, is "X rays." The "X" stood for "unknown" because Roentgen did not know what they were. At that time, physicists were trying to understand the merging of electric and magnetic phenomena, described theoretically by the Scottish physicist James Clerk Maxwell thirty years earlier. The theoretical physicist Hermann von Helmholtz, working from Maxwell's theory, had predicted the existence of "invisible high-frequency rays."[5] The rays would be invisible because the frequency was higher than the human eye could see. Von Helmholtz predicted that some of these invisible high-frequency rays would interact minimally with matter and so have penetrating power. Roentgen set out to find those rays.

On that Friday, Roentgen was late for dinner.[6] He had seen something eerie in his laboratory. His experimental setup consisted

Fig. 2.3. Wilhelm Roentgen, the discoverer of X rays and the first winner of a Nobel Prize in physics. (Image used by permission of the American Institute of Physics, Emilio Segrè Visual Archives, W. F. Meggers Collection.)

of cardboard surrounding a cathode-ray tube. The modern equivalent of a cathode-ray tube is a television picture tube;[7] X-rays are generated when the electron beam from the cathode slams into the anode target.[8] Roentgen investigated the properties of his new rays by placing various materials of varying thickness into the X-ray beam. He looked to see which materials allowed the beam to pass through and which materials blocked it. Because he knew the density and thickness of the materials he placed into the beam, Roentgen could infer the properties of the beam. One object he placed into the beam was a disk of lead. The lead blocked the beam completely, which in itself was an important clue toward understanding the nature of the rays. While placing the lead disk into the beam, Roentgen saw his hand; more specifically, he saw the *bones* of his hand. H. Seliger, writing about Roentgen's discovery in an article in the November 1995 issue of *Physics Today*, describes Roentgen as an "extremely reticent man" so that "one is forced to speculate about his views." Very likely, the apparition of the bones of his hand, floating before him in his darkened laboratory, startled him. No one had described similar apparitions in the physics literature of his day, yet several people were working with cathode-ray tubes and pursuing equivalent experiments. As the evening wore on, Roentgen apparently became increasingly confident that he was onto something new.

A bit of scientific context is important here. In 1895, we did not understand the atom. The electron had been discovered just a few years earlier. The basic theory describing the structure of the simplest atom in the universe, an atom of hydrogen, lay nearly 20 years in the future. The electromagnetic spectrum, although implicitly included in Maxwell's theory, was not understood as an entity in and of itself. That is why von Helmholtz wrote a paper describing the expected properties of the "invisible, high-frequency rays." Roentgen's X rays could be described, but they were not understood for quite some time. Occasionally, this is the way scientific knowledge advances: a discovery is made but is unexplainable by the scientific understanding of the day. Only after other scientists pursue different lines of research does the discovery take its place in the larger body of knowledge.

Roentgen continued his tests. He produced photographs of the phenomenon as proof. One photograph, of the bones of his wife's hand, including her ring, created a sensation when it was published in the January 5, 1896, edition of a Vienna newspaper. Subsequently, newspapers around the world picked up the story. The photographs were the 1896 equivalent of discovering a new planet or a new dinosaur; Roentgen promptly became a celebrity. Within four months, the American inventor Thomas Edison carried out his own experiments and placed an advertisement in the magazine *Electrical Engineer,* offering for sale "X-ray Apparatus of All Kinds for Professionals and Amateurs."

The photograph of the hand of Roentgen's wife not only communicated the power of the new physics, but also showed a valuable application in medicine.[9] As a result of his discovery of X rays, Roentgen won the first Nobel Prize in physics in 1901.

Roentgen used photographic plates to detect X rays from his experiments. Occasionally, astronomers still use plates for detection; more-modern methods of detection involve one of four methods: photographic film, proportional counters, charge-coupled devices (CCDs), and calorimeters.

When an astrophysicist says that he or she has detected an X ray, what does that statement mean? Exactly what is detected? To be detected, an X ray must interact with the measuring device or detector and produce some measurable effect. Ideally, the measurement will determine the specific quantities mentioned earlier: position, arrival time, and energy. As Roentgen showed, X rays are energetic, so they penetrate or pass through most materials. If, however, the material is very thick or made of matter with a high atomic number, the material may stop the X rays. That we say an X ray "passes through" a material is not quite accurate, because any material possesses the possibility of stopping an X ray. Any material attenuates a beam of X rays; the amount of attenuation depends on the atomic number of the material, its density, its thickness, and the energy of the X ray itself. The detection of an X ray is therefore a difficult problem because X rays interact with all materials through which they pass, including the very materials used to build the detector. This is one reason why every detector requires careful attention to calibration, particularly where the interaction of the wavelengths of light used with the detectors may compromise the observational goals of the project. The X-ray and gamma-ray bands are two such areas.

Nearly everyone is familiar with the X-rays used in dentists' and doctors' offices. There, the X rays have interacted with light-sensitive grains on photographic film.[10] The absorption of the X rays by the grains causes the grains to undergo a chemical reaction. Grains not chemically altered are removed during the processing of the film. The film records those X rays that have not passed through the teeth or bones; the X rays easily pass through skin and muscle. In other words, the dental or medical X ray is really the shadow of the teeth or bones, just as a shadow on a bright sunny day occurs because the presence of the person's body prevents photons of sunlight from reaching the ground. X-ray astronomers used this approach to see clouds of cold gas in the interstellar medium; X rays from sources beyond the cold cloud are absorbed in the cloud, creating a shadow in that direction. Wider application of this approach is impossible, however, because we would have to move a bright source of X rays around the universe.

The accumulation of photons is an observational goal of astronomy. Progress in astrophysical understanding is directly tied to how well we accumulate the light we receive. Our eyes do not accumulate light. Telescopes, first used by Galileo to look at the Moon and planets, collect more light than our eyes, so we see fainter objects, but a telescope and an eye still do not accumulate photons.

Photographic film and photographic plates replaced the eyeball at the telescope because film and plates accumulate photons. Photographic plates are similar to film, except the emulsion coats a sheet of glass rather than a plastic base. Previously, astronomers could only record what they saw by sketching the image or by counting stars. Astronomers with poor eye sensitivity were at a clear disadvantage.[11] Photographs of the sky quickly revealed many more stars than were visible even to a telescope-aided eye and recorded very faint emission from clouds of gas. Photographic plates were also rather stable, so the plates were stored for later study; stored plates were particularly valuable when studying time-variable objects.[12]

Photographic film has several drawbacks of which one is important here.[13] The process of creating an image is a photochemical one. Photographic film is really a layer, called an emulsion, of light-sensitive grains, usually one of the silver halides (silver chloride, silver bromide, or silver iodide), on a flexible plastic sheet. The photochemical process by which the grains capture light is actually quite complex. When light strikes the emulsion, some of it is absorbed by one or more grains. The light converts the silver halide into atoms of silver. Chemical processing of the film washes away the unexposed silver halide grains and fixes the silver to the plastic. From an astronomer's viewpoint, there are problems with this situation. The amount of light needed to alter the grains depends on their size. Large grains, which collect more light than small grains, are more likely to be altered. That means emulsions with large grains are more sensitive than those with small grains. But large grains produce fuzzier pictures. One grain can be completely converted to silver even though it may have been exposed to half the light of its neighbor, so the image can be fuzzy at the edges. It is difficult to manipulate film for maximum scientific return. Film can be scanned and digitized, but at the cost of introducing another layer of manipulation. In spite of these problems, photographic film and plates were used for more than 100 years in astronomical research simply because nothing else was available. Photographic plates were used until recently for surveys of large areas of the sky because plates can be produced inexpensively. The first electronic sky survey of a portion of the sky, the Sloan Digital Sky Survey,[14] has only recently become operational.

If photographic film is a relatively poor detector for astronomical purposes, what can we use? We need something that detects the X ray directly because we need to see the astronomical sources of X rays. Let's take a clue from Roentgen's experiments. Lead, for example, stops X rays: its atoms are sufficiently large (because of their high atomic number) and its density is sufficiently high that X rays cannot penetrate a thick slab of it. The intensity is reduced exponentially with the increasing thickness of the slab. Therefore, X rays will interact with a material if it is sufficiently dense or of a high atomic number. Rather than stopping the X ray completely, as with lead, we need to detect the X ray's interaction. Instead of using a solid, we will turn to a gas. We already know this works: the atmosphere of Earth stops X rays because otherwise we could detect X rays from the Sun on the ground. We will stop X rays if we build a box and confine the equivalent of one atmosphere of gas within it.

This is the interaction for which we have been searching. The X ray, upon interacting with an atom of the gas, will impart sufficient energy to one or more electrons to allow them to escape the electrical attraction of the atom's nucleus. The energy the electron carries away depends on the energy of the X ray. The ion will end up with a positive charge. To record that interaction, we introduce a mesh of wires, much like the screen in a window or screen door,[15] into the box. If we place a known, positive charge on the wires, the resulting electric field of the mesh will attract the liberated, negatively charged electron. When it reaches the wire, it will reduce the positive charge. The reduction of the charge will cause an electric current to flow to replenish the charge lost to the electron. The amount of charge that must be replaced will be equal to the amount of the electron's charge. The electronics record the location of the interaction through the disturbance of the electric field—in other words, through the flow of current. This description may be easier to picture using water. Imagine a grid of pipes sitting above buckets of water. Each bucket connects to the grid with a depth gauge and a spigot. One job exists for the depth gauge: maintain a constant amount of water in the buckets. If someone takes water out of any bucket, the depth gauge for that bucket will turn on the spigot for that bucket. Water will flow throughout the entire grid, but will only appear in the bucket with the open spigot. In this analogy, the water represents the electric current.

Fig. 2.4. A sketch of a proportional counter and the interaction that occurs within. The collimator is designed to prevent X rays from entering the counter from directions not toward the target. The crossed wires are embedded within the proportional counter and surrounded by the counter's gas. The interaction of the X ray and an atom of the counter gas ejects an electron from the atom. The ejected electron ionizes other atoms, creating a small electron cloud; approximately one electron-ion pair is created for every 30-keV of X-ray energy.

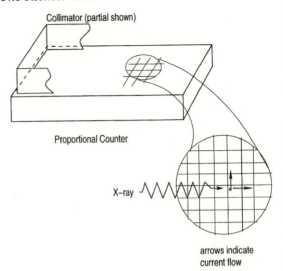

Collimator (partial shown)

Proportional Counter

X-ray

arrows indicate
current flow

Such a box filled with gas is called a proportional counter,[16] where the proportional part comes from tuning the voltages so that the output voltage is proportional to the amount of ionization that occurs in the gas. It works essentially this way, with one addition to the description above (Fig. 2.4). The counter must sit behind either a telescope or a collimator. A telescope focuses the light onto the detector; a collimator merely blocks X rays, except those that lie along the pointing direction, from reaching the detector. Think of a collimator as like your hand when you shade your eyes from the Sun or from glare. Collimators are less expensive to build than telescopes; early X-ray satellites made frequent use of them.

That description sounds great. Yet astrophysicists do not want new satellites to carry proportional counters. Why not? The issue is one of resolution: measuring the position requires a wire mesh. The finer the wire mesh, the better the spatial resolution. Take a close look at a window screen. If the wire mesh is too wide, mosquitoes will fly directly through the screen. You, the homeowner, want a tight mesh for your window screen to keep the insects outside where they belong. The tightest mesh is a solid piece of metal. Presumably, you also want the cool breeze to enter. That goal pushes you toward an open mesh.

In a proportional counter, the position of the X ray will be uncertain in inverse proportion to the size of the gaps in the wire mesh. We may not make the wire mesh too fine, however, because the wires will essentially lie next to each other, defeating the purpose of using a gas. Furthermore, if the wires lie too close to each other, the electric field surrounding a specific wire will interfere with the electric fields generated by nearby wires, distorting the signal induced by the X ray.

Physicists have used proportional counters for decades, for example, in recording the output of collisions between particle beams in a particle accelerator.[17] Astronomers adopted the technology for satellite observations. Proportional counters have two distinct advantages: it is easy to build a large counter to detect huge numbers of X rays, and large counters are rather inexpensive to build. In a later chapter, we will return to see the value of large proportional counters.

Until the late 1940s, X rays were believed to exist only in the laboratory because of the difficulty of creating them. Laboratory experiments established that air absorbs X rays, explaining the lack of X-ray detections on the ground. A rocket flight in 1949, carried out by researchers from the Naval Research Laboratory, firmly established the Sun as a source of X rays.[18] The Sun was, however, X-ray bright because of its proximity. If the Sun were as far away as some of our nearest stellar neighbors, the 1949 rocket flight would not have detected solar X rays. As a result, during the 1950s, most astronomers and physicists did not believe any nonsolar X-ray sources existed, so they left the nascent field to just a few persistent workers. Carrying out research in astronomy by rocket was, and still is, not for the faint of heart. An investment of years of hard work may return all of about five to ten minutes of data. Few on Wall Street would invest, given the apparent low return. Rockets are relatively inexpensive, however, so they were, and still are, perfect for testing new detectors and searching for other X-ray sources. If you know nothing, any information is considered valuable.

On June 18, 1962, at 11:59 P.M. (Mountain Standard Time), a team of physicists and astronomers fired off a rocket,[19] their third attempt aimed to detect X rays from the Moon. Their first attempt had exploded; during the second, the door covering the detector had failed to be jettisoned, so the detector was never exposed to the sky. The official goal of the experiment: the detection of X-ray fluorescence from the lunar soil as a means to identify the composition of the soil. In other words, X rays and charged particles from the Sun hit the Moon and cause the material that makes up the lunar soil to give off X rays. The detection of any X-ray source other than the Sun constituted the unofficial goal. The rocket, launched successfully, reached a maximum altitude of 225 kilometers. The experiment on one hand failed, but on the other hand succeeded. It failed because it did not detect the Moon. Not at all.

It succeeded because it discovered the first nonsolar X-ray emitters. The X-ray detector was sufficiently sensitive that it detected an overall X-ray glow covering the entire sky (Fig. 2.5 [see next page]). That glow is called the "diffuse X-ray background." The experiment also discovered a source of X rays, the large hump in Figure 2.5, in the constellation Scorpius (Fig. 2.6 [see next page]); that source became known as Sco X-1 (the first X ray source in Scorpius). This discovery would later be heralded as the birth of X-ray astronomy.

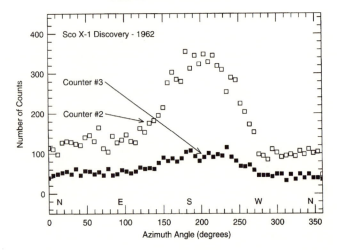

Fig. 2.5.   The data that led to the detection of Sco X-1 and the diffuse X-ray background from the 1962 discovery-rocket flight. Counter number 2 (open squares) recorded soft X rays, while counter number 3 (filled squares) recorded hard X rays. As the rocket rotated (the x-axis or azimuth angle), the detector scanned a strip across most of the sky, including the constellation Scorpius. The large increase and decrease of the open squares is the X-ray binary Sco X-1 coming into and passing out of the detector's view. Note that the background, when Sco X-1 is not in view (about 0 to 100 degrees and about 270 to 360 degrees), is higher on the left (east) side of the plot than on the right (west). This is the first evidence of X-ray emission from the galaxy itself or from a diffuse background contributor. This plot essentially started X-ray astronomy, so its information value is high, in spite of its locating Sco X-1 to within ±5 degrees. (Plot reproduced from the original and used by permission of Dr. R. Giacconi; the original appeared in *Physical Review Letters* 9 (1962): 439.)

Fig. 2.6.   The strip of sky scanned by the detector on the 1962 rocket flight. The nearly vertical line is the rocket's path. The detectors essentially looked at all parts of the sky above the rocket's horizon (curve marked "horizon"). Luckily, Sco X-1 is bright enough to be detected easily in the data. (Plot reproduced from the original and used by permission of Dr. R. Giacconi; the original appeared in *Physical Review Letters* 9 (1962): 439.)

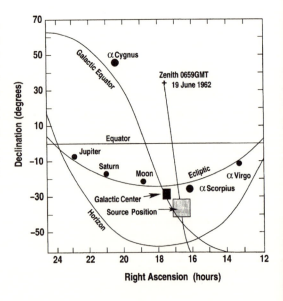

The 1962 discoveries remain impressive because rocket flights are measured in minutes above the atmosphere; that 1962 discovery flight lasted 5 minutes and 50 seconds above an altitude of 80 kilometers. This height is critical because above it,[20] the atmosphere does not absorb X rays, so a rocket-borne detector can begin to see the X-ray sky. Furthermore, to provide stability, the builders forced the rocket with its detector to spin, reducing the effective exposure time on a given part of the sky. If the instrument could detect nonsolar X-rays in less than six minutes, while spinning, the study of X rays from the universe might not be a quixotic venture; there really might be a few objects sufficiently bright to make the effort worthwhile.

Subsequent rocket flights supported that inference. A group from Lockheed started using a gas jet–controlled rocket, thereby increasing the pointing stability. This meant that the detector executed a slow scan, permitting sensitive searches for point sources. All the other X-ray groups implemented the technique. As a result, by the 1965–67 period, astronomers had collected about two dozen or so nonsolar X-ray sources. Included among them were clues that whetted appetites about the possibilities of interesting astrophysics. Among the finds, count the discovery of the transient source Centaurus X-2 (Cen X-2). Transient means that the source varies, but the connotation of the word is a rapid brightening and dimming. Cen X-2 increased its flux by more than a factor of 100 and then faded over an apparent period of a few months. Rapid variability generally implies a compact object because only compact objects can vary quickly. This line of reasoning is known as the "light travel time" argument; we'll encounter it in more detail later.

Second, the rocket scientists detected X rays from the Crab Nebula. During one of those early flights, they used the Moon, whose X rays still had not been detected, as the celestial equivalent of Roentgen's lead disk, studying the X rays from the Crab as the Moon crossed over the location of the nebula.[21] For this experiment, the lack of lunar X rays was good; the Moon served as an occulting disk. Two key points make this experiment important: the Moon moves at a known rate across the sky, and timing the disappearance and subsequent appearance of a source of X rays is easy to accomplish, even with rockets. The combination of known rate plus high time resolution provides a precise measure of the angular size. The data showed the Crab Nebula to be an extended X-ray source because the emission did not disappear in an instant as would occur if the entire Crab Nebula were a point of X rays. This, then, was a significant discovery.

With several X-ray sources known, a race ensued to improve the angular resolution of the detectors sufficiently to determine increasingly precise locations for them. Astronomers would then follow the X-ray detections with radio or optical observations to identify counterparts. These follow-ups were important for all the objects that did not fall along the Moon's path. By 1966, a rocket team led by H. Gursky reduced the search box around Sco X-1 by about a factor of a hundred (Fig. 2.7 [see next page]),[22] creating an opportunity to identify the optical counterpart.

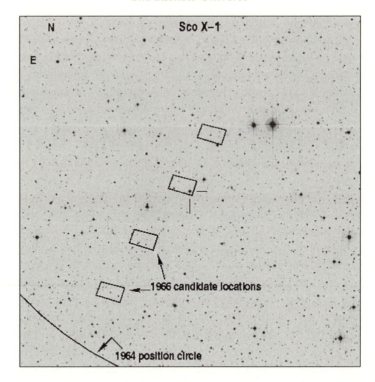

Fig. 2.7.   The phrase "looking for a needle in a haystack" illustrated: the image is about one-twentieth of the region surrounding the proposed location of Sco X-1 based on the 1962 rocket flight. Subsequent flights in 1964 and 1966 narrowed the possible counterparts for Sco X-1 to an object in this image. An arc of the full position circle from the 1964 flight is visible in the lower left corner. The small rectangular shapes result from the 1966 observation. The optical counterpart could lie within any one of the boxes, but the middle two were statistically favored. The bright star in the lower right box near the center, as indicated by the short lines, is the optical counterpart of the X-ray source. (The background optical image was obtained from the Digitized Sky Survey available at the Space Telescope Science Institute; see the Additional Figure Credits on p. 198 that are included here by reference.)

An improved position quickly led to an optical follow-up and an identification of a candidate counterpart, with direct implications for the development of X-ray astronomy. Allan Sandage from Palomar Observatories located a very blue star with an unusual spectrum near the position of Sco X-1. This star represented a significant connection between X-ray and optical astronomy because, as silly as it may sound, it was not the Sun and it was not the Crab Nebula. It represented a new type of X-ray source.

Why was it crucial to link X-ray and optical astronomy? The amount of information available to an X-ray astrophysicist was, and still is, considerably less than the information available to an astrophysicist working in the optical band simply because optical telescopes and instruments have studied many more objects.[23] Following the demise of the Einstein Observatory in 1981, the X-ray catalog of known objects contained about five thousand sources. That number did not increase until the launch of ROSAT in 1990, which detected about one hundred thousand X-ray sources. The increased size flowed from the all-sky survey ROSAT carried out during its first six months in orbit. Many of the cataloged objects are sufficiently faint that they cannot be studied in detail. For comparison, optical catalogs contain about ten million objects.

During the first few years (from 1962 to 1966 or so), X-ray astronomers essentially knew each object intimately because they could nearly count the number of known sources using their fingers. No one understood in detail how the X rays were produced because the accuracy with which the position of the X-ray source could be measured was poor. Without the connection of X rays to an optical source, thoughts on the nature of the sources and the X-ray-emission mechanisms largely went nowhere. Theories abounded, but science makes little progress without accurate data with which to test specific theoretical predictions. The connection to the larger body of knowledge of optical astronomy was crucial to understand what these objects were and how common they might be in the universe. Without that connection, X-ray astronomy would die like a tomato left on the vine. As a result, scientists tried hard to match the objects detected in the X-ray band with likely optical counterparts, as A. Sandage had done for Sco X-1. (We'll learn, in the next chapter, why matching X-ray sources to optical counterparts can be a difficult job.)

The three discoveries described above revealed a transient source (Cen X-2), a site where a star had died (the Crab Nebula), and a binary star (Sco X-1). These discoveries suggested that the X-ray universe was different from the relatively constant visible one. Excitement brewed as astronomers realized that the study of X rays would lead to a better understanding of their sources. That excitement motivated them to undertake the difficult work of designing and building sensitive instruments with which to study the X-ray universe. Relatively quickly, X-ray astronomers opened a new window to a different place: the energetic universe.

What happened to X-ray astronomy after those early rocket flights? Astronomers launched additional rockets, discovering about 30 sources. They also built "tagalong" instruments: an X-ray detector added to the side of a satellite that had a different mission. The tagalong instrument would detect X rays while the rest of the satellite carried out its mission. For example, the Orbiting Solar Observatory missions,[24] which spent their time looking at the Sun, carried tagalong X-ray detectors. So did the Vela missions. The Vela satellites were built by scientists at the Los Alamos National Laboratory to detect detonations of nuclear weapons. These satellites are, luckily for X-ray astronomers and the rest of us, much better known for what we learned from their astrophysics discoveries. Gamma-ray bursters (GRBs), brief but intense bursts of gamma rays, were discovered and demonstrated to lie beyond the solar system. Long a complete mystery because of the apparent lack of emission at any other wavelength, GRBs have become a hot target of observatories in the past few years. (GRBs will be discussed in more detail in a later chapter.)

The Vela satellites were built to remain in orbit for many years, so observations accumulated. The long orbit provided a few years of uninterrupted observing; the results of Vela 5B proved particularly interesting.[25] A few of the variable sources discovered, including the candidate black-hole system Cygnus X-1, varied repeatedly. Repetition in any science means that a pattern of behavior exists; patterns of behavior in nature attract scientists like a porch light attracts moths. Repetition means that the behavior can be observed and understood, and a model can be created to predict behavior, furthering the need for additional observations. This is scientific progress.

The first satellite dedicated to X-ray astronomy, launched in 1970 from Kenya, was called Uhuru.[26] It had the "large" collecting area of 0.084 square meters, which translates to a detector about 12 inches (30 centimeters) on a side. There are now amateur astronomers with telescopes that have a larger collecting area. In 1970, however, this was the state of the art. Uhuru was a spinning, or scanning, satellite (Fig. 2.8 illustrates the similar satellite Ariel V); it spun on its axis once every 12 minutes. A free-floating satellite must be stabilized if it is to be of use; spinning a satellite is the easiest way to stabilize it because the spin immediately establishes a direction in space. A detector, mounted on the side of the satellite, then sweeps across a large portion of the sky, collecting X rays as the satellite spins.

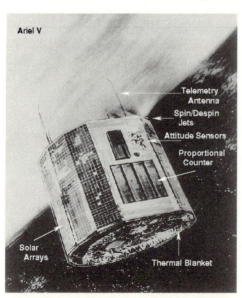

Ariel V

Telemetry Antenna

Spin/Despin Jets

Attitude Sensors

Proportional Counter

Solar Arrays

Thermal Blanket

Fig. 2.8. A sketch of Ariel V, an early X-ray satellite stabilized by spinning. Note the X-ray detectors that lie on the curved side of the spacecraft, so they sweep across the sky as the satellite spins. (Image obtained from the public archives at the High Energy Astrophysics Science Archive Research Center, NASA-Goddard Space Flight Center.)

This is the easiest way to survey the sky completely, because it covers the largest area in the shortest amount of time. A price is exacted, however. Detailed studies of a given object are more difficult to obtain with a spinning satellite because the detector looks in a given direction for only a short interval. Bright sources are detectable, but not faint ones. The position of an object in the sky must be reconstructed by recording the aspect of the satellite—the direction in which it is pointed and the orientation of the detector with respect to that direction—at all times. The detector's orientation is more difficult to establish with a spinning satellite; this imprecision limits the reconstruction by introducing an uncertainty in the direction from which the X rays come. That uncertainty translates into a poor position for, and a blurring of, the source. For a point source, such as a hot star, this blurring is irrelevant, provided the observer knows beforehand that the source is a point of emission. For an extended source, the blurring prevents the observation of any details.

Look at the view out of a side window of a car as the car drives down the road at a high rate of speed.[27] If you focus on a particular object, you must turn your head with the direction of motion to freeze its appearance on the same spot on your retina. If you do not, the object blurs. Scanning satellites fundamentally do not follow the apparent motion of an object, so everything they detect has a blur introduced by the rotation of the satellite. Returning to your window view, note that you get only a quick glance at a specific object as it first enters and then exits your view. You'll notice big objects, or bright ones, or ones that are isolated in some way. Small objects or objects surrounded by others will be blurred. This explains why faint X-ray sources cannot be detected. With a short spin period, any given point in the sky will be in view for only a few seconds. The total exposure time becomes the sum of many few-second glances. For bright sources, this approach is valuable because, over a long interval of time, the data may reveal the intrinsic variability of the source. Faint sources, however, will remain invisible.

The spinners provide one advantage: undertaking an all-sky survey is relatively easy and, if the survey is the first one, it becomes valuable. Uhuru detected 339 sources of X-ray emission, including binary stars, active galaxies, clusters of galaxies, and supernova remnants. More sources were discovered to be variable, establishing variability as a common property of X-ray astronomy. One of the key discoveries of Uhuru involved clusters of galaxies.

In 1970, E. Kellogg, H. Gursky, and their colleagues, then working at American Science and Engineering in Cambridge, Massachusetts, used Uhuru to study two clusters of galaxies in the constellations Virgo and Coma. They found a surprise.

What is a cluster of galaxies? Our solar system is gravitationally bound to the Sun: where it goes, the planets follow. In other words, the solar system moves as a unit. A galaxy is a collection of stars, gas, dust, and planets all moving through space together. A cluster of galaxies, then, is a gravitationally bound group of galaxies; usually, astronomers carve a distinction between a cluster, with more than 50 galaxies, and a group, with fewer.

The differentiation into groups and voids suggests that the formation of these structures occurred during the early development of the universe. A debate on the top-down versus bottom-up development of the structure has been waged for years. The top-down position implies that galaxies formed last; the bottom-up approach argues that galaxies formed first, then captured and collided to form larger objects such as groups and clusters. The consequences of the two approaches for well-studied objects such as galaxies are substantial. In the top-down approach, the material to make a cluster becomes isolated; galaxies must subsequently form by dissipating energy from knots in the isolated gas globules. In the bottom-up approach, galaxies can form as small knots in the expanding primordial matter. As theorists investigated the details of the early development of the universe, they realized that small perturbations survive under certain conditions of the universe, providing a theoretical foundation for the bottom-up approach. Unfortunately, additional work showed that large perturbations also survive under certain conditions. The conflict of the positions drives the desire for studies of ever-fainter objects increasingly farther away and consequently further back in time. Observations of objects across a broad range of the history of the universe must settle the debate. Observers pursued the necessary data, and since about the mid-1980s, the observational evidence favored the bottom-up approach.[28]

By itself, that clusters of galaxies emit X rays was not surprising because the individual galaxies that make up a cluster are X-ray sources. Why? We know that galaxies are made of binary stars and supernova remnants (such as the Crab), both of which are X-ray sources. Galaxies typically contain 10 million to 100 billion stars or more. Only a small fraction need be X-ray-emitting binary stars or remnants for a galaxy to be X-ray bright.[29]

Kellogg and his collaborators discovered that clusters possessed a large, hot halo of gas. In the optical band, a cluster appears to be a nonrandom grouping of galaxies (Fig. 2.9). Clusters are difficult to find at optical wavelengths because one must look at a rather large piece of the sky, peer past all the foreground stars, and pick out the clusters. All these objects are projected onto the two-dimensional image of the sky. By careful study, however, you find regions of space where many more galaxies than the average exist. Picture a cocktail party: as you enter the room, you can tell at a glance approximately how many people are present, whether 10, 100, or 1,000. What you see, however, are groups of people, some groups large, some small. There are relatively well defined groups as well as empty areas. The same description holds for galaxies. With work, the boundaries of a cluster observed in the optical band become visible.

In the X-ray band, however, clusters are stunningly obvious. The halo of X-ray emission essentially defines the size of the cluster (Fig. 2.10). The X-ray emission comes from hot gas gravitationally bound to the cluster. The gas is visible only in the X-ray band because it is hot, about 25 million degrees.[30] Whereas the optical definition of a cluster is a well-separated grouping, in the X-ray band, the definition of a cluster becomes "a collection of galaxies embedded within a hot bubble." Think of a tomato: the entire tomato represents the hot-gas bubble; the tomato seeds represent the individual galaxies. If you place the tomato on the ground and spread some seeds near it, which is easier to spot: a few seeds, or the entire tomato?

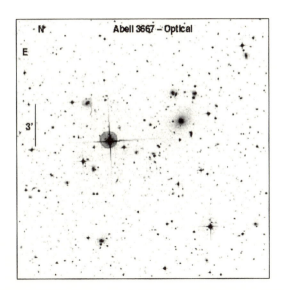

Fig. 2.9.   An optical image of a cluster of galaxies, here, the cluster Abell 3667 (the 3,667th entry in a catalog of clusters by G. Abell). Without a pointer to the cluster, its location is not apparent. However, the majority of fuzzy objects near the center of the image belongs to Abell 3667. (The optical image was obtained from the Digitized Sky Survey available at the Space Telescope Science Institute; see the Additional Figure Credits on p. 198 that are included here by reference.)

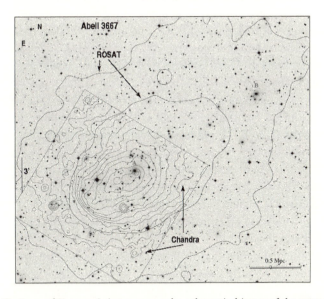

Fig. 2.10.   Contours of X-ray emission superposed on the optical image of the same cluster. The contours connect regions of equal X-ray emission. Two sets of contours are presented: the large, widely spaced contours are those from a ROSAT observation; the tighter contours are those from a recent Chandra observation. The actual X-ray image of Abell 3667 is presented in chapter 8. (Image used by permission of Dr. Alexey Vikhlinin, Harvard-Smithsonian Center for Astrophysics; the original image appeared in *Astrophysical Journal* 551 (2001): 160. The background optical image was obtained from the Digitized Sky Survey available at the Space Telescope Science Institute; see the Additional Figure Credits on p. 198 that are included here by reference.)

The hot gas helped address a thorny issue first raised in the 1930s in research on clusters of galaxies. When astronomers measured the speeds of individual galaxies in various clusters, the values appeared to be too high for the clusters to remain gravitationally bound. When we studied (optically) distant clusters, they resembled nearby ones, suggesting that clusters live relatively unaltered for a long time. This was a problem.

What prevents the galaxies from speeding away from one another? The motions of the individual galaxies lead to an estimate of the total mass of the cluster. The sum of the masses of the individual galaxies constitutes about 5 percent of the total mass. The hot gas observed in the X-ray band makes up part of that unseen matter: it represents about 20 percent of the mass needed to keep clusters together. The existence of the hot gas in clusters is one reason why X-ray astrophysics is so important. The total mass of hot gas is greater than the sum of the masses of the individual galaxies, just as the tomato contains more mass than its seeds. If you deprive yourself of an X-ray image of a cluster, you will never understand its structure or formation because you will be missing one of the most important pieces of evidence.

What constitutes the other 75 percent? A good question; the answer is that, right now, no one knows. Intrepid researchers postulated the existence of unseen matter. Why? The mass must be there because of its gravitational effect on the constituent galaxies. It must also be undetected, at least to date. The possible explanations all describe "dark matter," dark because we cannot see it and matter because we know, from its gravitational effects, that it must be present.

This is a not small problem. Galaxies themselves must be embedded in a halo of dark matter. That result was first obtained in the 1930s by Fritz Zwicky after he measured the rates of rotation of galaxies and found they were rotating faster than expected. The rate of rotation depends directly on the mass of the galaxy, so faster rotation meant galaxies had more mass. Astronomers computed expected masses based on the amount of light emitted: if every object emits light in direct proportion to its mass, then by adding up the amount of light from a galaxy, its mass may be estimated. The flaw in the analysis is this assumption: if galaxies spin faster because their mass is higher, then not every object emits light in proportion to its mass; hence, "dark matter." Estimates of the fraction of dark matter in a galaxy range as high as 95 percent.

Potential candidates for the dark matter exist: exotic new particles as yet undiscovered by particle physicists, generically known as "WIMPs," for Weakly Interacting Massive Particles; neutrinos with mass; or simply very faint objects. WIMPs by definition will be difficult to detect. If new particles remain unfound, searching for them will also be difficult, because there is little evidence to constrain the search.[31] Attention has focused on the other two possibilities.

Particle physics must now confront the evidence that neutrinos have mass. The idea for massive neutrinos may arguably be attributed to astronomers. The German physicist Wolfgang Pauli first postulated the existence of neutrinos in 1930 to balance the equations of nuclear transformations such as radioactive decay. In the decay of a neutron into a proton and an electron, for example, a small quantity of energy appears to evaporate if the neutrino is not included. Conservation of energy is a consistent, reliable behavior of the universe (a law of physics); physicists would fight long and hard for it before accepting its downfall. If we assign the apparently evaporating energy to the neutrino, the equation again balances. The conservation of energy is thus saved at the expense of postulating an unseen particle. Confirmation of the existence of the neutrino occurred in the 1950s.

Nuclear reactions occurring at the center of the Sun emit neutrinos.[32] Astronomers, particularly Dr. John Bahcall of Princeton University, emphasized that the observed rates of neutrinos emitted by the Sun failed to match the predictions of the theoretical models of the Sun by about a factor of three.[33] Particle physicists considered the arguments seriously only after Bahcall and his coworkers rigorously examined the details of the solar models. No matter how they twisted the models, Bahcall and his coworkers could not make the theory predict as few neutrinos as were observed. Nor was the shortage caused by a failure to detect the neutrinos. The observational apparatus was thoroughly tested and shown to detect neutrinos accurately.[34] There just were too few neutrinos. Arguments ensued.

One critic noted that the detected neutrinos had originated from one of the high-energy reactions occurring in the solar core. Those reactions are very temperature sensitive, so a slight temperature decrease at the core reduces the expected number and solves the problem. More twisting and pulling on the models, however, did not produce the expected change. Attention gradually turned to neutrino behavior. If, as proposed in the late 1970s, neutrinos somehow switched their identity before detection, then the numbers might be reconciled. Neutrinos may switch their identities, or oscillate, but they must have a very small, but nonzero, mass. The oscillation then results from the interaction of the "massive" neutrino with the material along its path of propagation as it leaves the Sun's core.

Physicists became interested in the experiments as a path toward a greater understanding of the physical properties of neutrinos. They built additional experiments, all verifying the paucity of neutrinos from the Sun.[35] Physicists and astronomers eagerly anticipated detailed results from the Sudbury Neutrino Observatory (SNO), in Sudbury, Ontario, because it is designed to test additional components of neutrino theory. SNO physicists announced the first results in June 2001; they confirmed the paucity of neutrinos and concluded that neutrinos oscillate from one state to another between their creation in the core of the Sun and their detection in the heavy water in the SNO instruments. We do not know whether the oscillation occurs in free space or is catalyzed by the passage of neutrinos through the Sun's matter. Regardless, the neutrino experiments measure the interior temperature of the Sun's core: 15.7 million degrees, to within about 1 percent accuracy.[36]

Not to be outdone by researchers using abbreviations that lacked testosterone, astronomers first suggested that very faint stars or large "Jupiters" could constitute much of the dark matter. The abbreviation coined for this explanation: MACHOs (MAssive Compact Halo Objects).[37] In this case, the objects would not really be dark, but simply not yet detected, dumping most of their energy into the mid- or far-infrared bands where observations remain scanty and the background is high. Recently, astronomers from the University of California at Berkeley and the University of Edinburgh identified about three dozen objects in the halo of the Milky Way that they argue are in fact faint, cool white dwarfs.[38] They showed that these white dwarfs contribute perhaps as much as 35 percent of the dark matter in galaxy halos. Although their results require confirmation, their study also benefits researchers interested in the formation of spiral galaxies because astronomers expect halos to form before the disk.[39] White dwarfs live for a long time, so the objects discovered in the halo of the Milky Way may well be among the first stars to shine from the collapsing cloud of gas that became our galaxy.

Before we leave the subject of galaxy clusters, there are two items to consider. First, the alert reader should be asking a question: how do we know that the bubble of gas in a cluster is hot? The answer to that question will come later, after we learn spectroscopy. Second, the discussion in the previous paragraphs may obscure the implications of dark matter. Clusters of galaxies contain most of the luminous matter in the universe; that matter is hot gas. Clusters also contain most of the dark matter in the universe, whatever it turns out to be. That means that matter much unlike us constitutes an overwhelming part of the universe in which we live, and matter that is like us is in a form unlike what we generally encounter.

The X-ray detectors and missions up to the mid-1970s were either tagalong detectors or spinning satellites. To make progress, X-ray astronomers needed observatories, so they could image a specific target for an exposure time determined by the scientific goals of the observer and the need to achieve a given sensitivity. With pointed observations, fainter sources immediately become detectable because the accumulation of X rays gradually exceeds the random arrival of background X rays. X-ray astronomy needed a satellite that imaged X-rays.

In the late 1970s, NASA launched a series of large satellites dedicated to high-energy astrophysics. The first, HEAO-1 (for High-Energy Astronomy Observatory), was a spinner. The difference from previous missions was its size: it carried detectors with collecting areas larger by a factor of 10 over previous missions. It carried out an all-sky survey, detecting about 840 X-ray sources scattered across the sky. There are 41,255 square degrees of sky; if the HEAO-1 sources were distributed evenly, there would be one source for every 49 square degrees.[40] Optical catalogs contain tens of millions of sources; distributed uniformly and assuming 50 million sources, we expect one source for about every 0.0008 square degrees. The difference in average sky density may convey a sense of the eagerness with which X-ray astronomers in the mid- and late 1960s wanted more-sensitive telescopes.

The second satellite in the HEAO series was renamed Einstein after launch because it was placed into orbit in the hundredth-anniversary year of the birth of Albert Einstein. It was the first imaging mission that detected X rays after focusing them. New areas of X-ray astronomy immediately opened with data from Einstein. Normal stars, those much like the Sun, not only emitted X rays but did so in great numbers. Elliptical galaxies, believed to be devoid of gas, implying that new stars could not form, instead were found to contain apparently hot, diffuse gas similar to a hot interstellar medium. Some of the gas appeared to be flowing away from the galaxies' nuclei. A series of long observations of a small patch of sky (780 square degrees, less than 2 percent of the sky) detected 835 objects; if this patch of sky were typical, then a complete sky survey to the same level of sensitivity might detect 44,000 objects.[41] The key contributions of Einstein were its images. For the first time, we saw how the universe would look if we had X-ray-sensitive eyes.

Einstein fell silent in April 1981 after observing about five thousand X-ray sources. That date represented the start of what was to be a long dry spell during which no additional observations would be obtained with American-built X-ray satellites. Chandra existed on paper; it was in its earliest planning stages.

During the 1980s, the Japanese started their X-ray astronomy program, launching a series of increasingly sophisticated satellites. The first was Hakucho (in English, Swan; the renaming of a Japanese satellite is an honor bestowed by the launch team). Although Hakucho was a spinner, its instruments were oriented to look along the spin axis, so it functioned much like a pointer. Hakucho detected several new burst sources and uncovered new behaviors in several well-studied sources. The second satellite, launched in 1983 and named Tenma (Pegasus), was a larger version of Hakucho. Tenma found emission lines of iron in several X-ray binaries; this was the first time lines were observed in binaries. (Spectroscopy is discussed in chapter 6.) The third satellite, Ginga (Galaxy), roared into orbit in February 1987, just in time to turn its large proportional counters, the largest lofted to that date, on a supernova that blew up in our galaxy's nearest neighbor, the Large Magellanic Cloud (Supernova 1987A). The fourth satellite of Japan's X-ray astronomy program, ASCA, is described in chapter 6. The ill-fated Astro-E was to have been the fifth.

We know how to obtain the positions of sources of X rays. What can we do with that knowledge? First, we compare the X-ray positions with positions of objects listed in catalogs compiled from observations in the optical, radio, ultraviolet, and infrared bands, and search for matches.

Many X-ray sources, detected by Einstein, match up with single stars in our galaxy. Astronomers classify stars by their surface temperature (from hot, about 50,000 degrees K, to cool, about 3,000 degrees K) and their luminosity or their size. A relation exists among luminosity, temperature, and size, so knowledge of any two quantities is sufficient to calculate the third.[42] The Sun is average, with a temperature of about 6,000 degrees K and a luminosity of 2×10^{33} ergs per second.[43] Define that value to be one solar luminosity; we'll then measure all luminosities relative to the Sun. The Sun is also an X-ray source. Because the Sun is average, we predict that hot, single stars will be bright X-ray sources while cool, single stars will be faint. What do the data show?

The data reveal that, in general, cool, single stars are brighter X-ray sources. Puzzling? You bet. As it turns out, even the hottest star is not sufficiently hot at its surface to be an X-ray source. So how can cool, single stars be X-ray bright? To answer this question, we must first understand spectroscopy. We'll get there shortly and then return to this puzzle.

What other objects in our catalogs have been found to emit X rays? Astronomers discovered some X-ray sources to be extended—that is, the X rays were not emitted from a single location but were distributed across some portion of the sky. On occasion, an extended source in the X-ray band corresponded with an extended optical or radio source. The Crab Nebula, the remains of the stellar explosion observed by humans in A.D. 1054, is an extended source in every wave band studied to date. So, too, are most nearby galaxies.

Among the first objects examined by Chandra during its commissioning activities was the active galaxy PKS 0637-75. ("PKS" stands for ParKes Survey, a radio survey conducted by the Parkes Radio Telescope in Australia.) The Chandra mirror team expected PKS 0637-75 to be a point source as seen with Chandra's high-

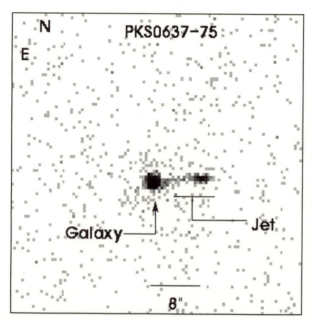

Fig. 2.11. The active galaxy PKS 0637-75 as observed by Chandra during its commissioning phase; the galaxy was one of the first targets acquired. It was chosen because it was believed to be a bright point source and would serve as an excellent test of the quality of the mirrors. Instead, the observation yielded the largest and brightest X-ray jet yet observed. (Image generated from data obtained from the Chandra public data archive at the Smithsonian Astrophysical Observatory. Similar figure appeared in D. Schwartz et al., *Astrophysical Journal* 540 [2000]: 69.)

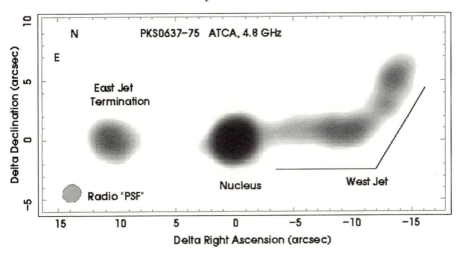

Fig. 2.12. The active galaxy PKS 0637-75 as observed by a radio observatory in Australia, the Australian Telescope Compact Array. The radio jet corresponds precisely in position to the X-ray jet observed by Chandra. (3cm [4.8 GHz] image of PSK 0637-752 made with the Australia Telescope Compact Array courtesy Jim Lovell, CSIRO Australia Telescope National Facility.)

quality mirrors. Figure 2.11 shows the scientifically interesting result: it is instead a double source. Comparing the X-ray data to a radio image (Fig. 2.12) shows, however, that the two sources are related. The bright one is the nucleus of the active galaxy; it emits about 10^{11} times as much energy as the Sun, and does so from a region smaller than our solar system.[44] The fainter source is a jet of matter and radiation ejected from the central regions of the active galaxy. The jet had been previously detected in the radio; the existence of the X-ray jet was new, because no previous X-ray observatory had separated it from the nucleus. The observed length of the jet is about 100 kiloparsecs, or about 325,000 light-years. Radio observations reveal individual hot spots in the jet moving away from the galaxy nucleus at an apparent speed of almost 18 times the speed of light. Observations of apparent motion at speeds faster than light have existed for decades. It is not real motion, but rather a projection of motion onto the plane of the sky (in other words, across, or perpendicular to, a line from us to the galaxy nucleus).[45] From the measured expansion, we estimate the angle of the real motion to the plane of the sky; the result shows the jet points toward us to within about six degrees. Using that angle, the actual length of the jet is not 100 kiloparsecs, but closer to 1,000 kiloparsecs. It is the largest and brightest X-ray jet known. Chandra discovered it as its first observation of a celestial target.

Let's return to our match-up of X-ray sources and catalogs. Not every source shows an obvious match. Some sources show no match; others correspond to many matches. Why? If the position is poorly determined, then you, the astronomer, are "confused": which optical source corresponds to the X-ray source? The word "confused" is a technical term in this context. When attempting to match sources in different wave bands, you may be lucky: if the optical source is isolated, or is peculiar in some manner, you can demonstrate a match. That is how many optical-to-X-ray correspondences were uncovered in the early years. But when you do not know which source in band A corresponds to a source in band B, you have source confusion. Source confusion is an ugly situation that occurs whenever the resolution of the detector is broader than the average separation between celestial objects in a region of the sky. It is ugly because the interpretation of the data ranges from difficult to impossible.

The X-ray observations of the active galaxy NGC 6814 were source-confused for about 20 years. Active galaxies contain substantially more luminous nuclei in the X-ray band (and often other bands as well) than normal galaxies. Current models assume a galactic-mass black hole at the center of the galaxy.[46] A black hole has a radius, the Schwarzschild radius, defined as the distance interior to which light cannot escape the gravitational force of the hole. The Schwarzschild radius is directly proportional to the mass of the black hole.[47] Astronomers distinguish between stellar-mass and galactic-mass black holes. If they exist, stellar black holes form during the collapse of massive stars. The mass involved might be 5 to 10 solar masses, for example, giving a Schwarzschild radius of about 8 to 14 kilometers. The mass at the center of active galaxies is estimated to be about 5×10^7 to $\sim 10^9$ times the mass of the Sun.[48] For the estimated mass range, the Schwarzschild radius is approximately 0.3 to 20 times the size of Earth's orbit. The radius estimates are confirmed by variability arguments, in particular the oft-used "light-travel-time" argument.

The gist of the argument is as follows. Imagine a variable object: the flashing strobes of a police car. Most police cars have multiple strobe units in the roof rack; when the lights are active, each strobe emits a pulse of light, but they are not coordinated,[49] so a chaotic sequence of flashes occurs. If you wanted to coordinate the strobes, a signal would have to travel to all of the strobes to fire. That signal would have to arrive simultaneously at all strobes; otherwise you would not coordinate the strobes to flash in unison. You could easily send a signal to the nearest strobe, but the end units lie sufficiently far away that they would not necessarily receive the signal to fire at precisely the same time as the nearest strobe. And if the signals do not arrive at all strobes simultaneously, chaos reigns. This is the essence of the light-travel-time argument. If you observe a variation, then the mere existence of the variation argues for coordination and sets a limit on the size of the emitter: it must be smaller than the time required for light to travel from one side to the other. For a police car's strobe unit, this is easily achieved because the unit is itself only a few feet in length and the speed of light is great. Imagine, however, inflating the strobe unit to cover Earth; coordinating the units would be considerably more difficult.

Using the light-travel-time argument, we can estimate the size of a variable object, particularly when little other information is available. Variations in the X-ray band set limits on the size of the central engine of active galaxies. NGC 6814 was for years a unique active galaxy: astronomers uncovered a remarkably stable X-ray period of about 12,000 seconds (usually written as "12 ksec" or "12 ks"). A variation of length 12 ksec, following the light-travel-time argument, translates to an object smaller than about one-eighth of a light-day, the distance light travels in one day,[50] or approximately equal to the distance from the Sun to Uranus.

Data from several satellites yielded the same period, suggesting a robust result, but all the satellites used had limited spatial resolution. A relatively stable period in an active galaxy motivated researchers to consider various types of objects that might orbit the central black hole. Figure 2.13 shows the ROSAT observation of NGC 6814, which lies at the center of the image. Included in the figure are boxes and circles overplotted to show the sizes of the spatial resolution of several previous X-ray sat-

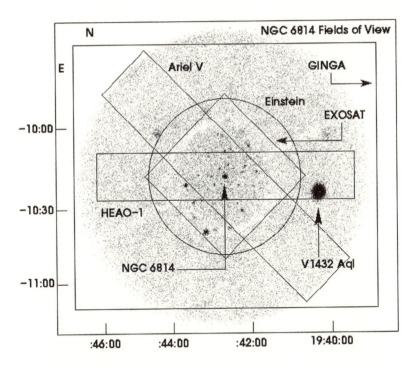

Fig. 2.13.   The active galaxy NGC 6814 as seen by ROSAT. The boxes indicate the spatial fields of view of the satellites (labeled), most of which used nonimaging proportional counters. The large circle indicates the field of view of the Einstein Observatory. The fields of view have not been oriented correctly; instead, all are centered on NGC 6814 solely to illustrate the concept of source confusion. NGC 6814 is the source at the center. Note that the large rectangles all cover the source to the west, but it falls outside the field of view of Einstein. The western source is the interacting binary star V1432 Aquilae, discovered by ROSAT from the data shown here. (Image generated from data obtained from the ROSAT data archive, NASA-Goddard Space Flight Center.)

ellites, most of which used proportional counters. The actual positions and orientations of the boxes are arbitrary; they were chosen solely to illustrate the nature of the problem. The resolutions were sufficiently poor that researchers identified the X-ray source in this part of the sky as NGC 6814 (Fig. 2.14). The confusion is apparent: the optical image shows an obvious bright source and a known active galaxy. What would you do? Likely, just what the astronomers analyzing these data did: identify the X-ray source in that direction as NGC 6814. As the ROSAT data show, the real source of the periodic X-rays is the bright object west of NGC 6814. This object happens to be an interacting binary star, specifically, a magnetic cataclysmic variable (V1432 Aquilae) with an orbital period of slightly more than 12 ksec. The X-ray properties of NGC 6814 itself, once disentangled from those of the binary, turned out to be quite ordinary for active galaxies.

NGC 6814 represents a perfect example of the need for imaging telescopes so that you not only know where, but also at what, you are looking. In addition to looking, however, you also need to use a high-quality instrument. In the next chapter, we extend some of the concepts discussed here to include the quality of the point-spread function.

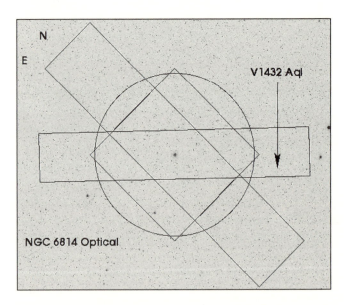

Fig. 2.14. An optical image covering the same portion of the sky as the previous figure. NGC 6814 is again at the center. The small circle to the right indicates the location of V1432 Aquilae. Given the existence of an X-ray source in this part of the sky plus the known active galaxy NGC 6814, it is completely understandable that astronomers identified NGC 6814 as the X-ray source. The identification influenced research on active galaxies for about 5 to 10 years before the confusion of sources was uncovered. (The background optical image was obtained from the Digitized Sky Survey available at the Space Telescope Science Institute; see the Additional Figure Credits on p. 198 that are included here by reference.)

3

"... every solution serves only to sharpen the problem, to show us more clearly what we are up against ..."

—Eric Hoffer

French painter Georges Seurat rebelled against the increasingly "nebulous illusions" of the impressionists.[1] Instead, he thought the light from a scene to be painted should be decomposed into dots of color "scientifically."[2] Although the pointillist technique had first been used by Henri Matisse in his *Luxe, Calme, et Volupté*, pointillism is most often identified with Seurat. His painting *A Sunday Afternoon on the Island of La Grande Jatte* reveals the technique: little dots of color distributed on the canvas. Imagine how this painting was carried out: Seurat, using his brush, dabbed a bit of color onto the canvas, moved his hand slightly, and dabbed again.[3] Using this technique, he gradually built up an entire image by creating a pile of color in all the areas that shared a common color.

La Grande Jatte appears fuzzy and out of focus. Seurat, in decomposing any particular color into minute dots, fuzzed the edges of all his subjects. In essence, he imposed a "detector function" onto the scene he painted using a somewhat poor "point-spread function."

Whatever do we mean? In a perfect world, an instrument would detect every photon perfectly and not leave any evidence of the process of detection. Unfortunately, we do not live in a perfect world. Our instruments, in the process of detecting the very photons we use to study an astrophysical source, also impress upon the data we obtain some signature of the instrument itself. To understand what that signature does to our data, we must calibrate the instrument. For scientific equipment, the accuracy of the calibration sets a limit to the precision of the measurements. The calibration of Chandra was a large effort taking many people-years of work. Let's postpone for now a discussion of Chandra's calibration; here, we just need to acknowledge that instruments distort data.

Calibration is not an esoteric concern without importance to the real world. Calibration is everywhere, and an "instrumental signature" exists for nearly every interaction we have with the world. Some of these signatures are more blatant than others. Police radar guns have been known to suffer from a lack of calibration, as evidenced by challenges in traffic courts. To be accurate, a radar gun must be able to measure the speed of a car to within one mile per hour. When we pump gas into a car, we want a gallon to be a gallon; we certainly would not agree to pay for more gas than the amount we pumped. Fuel gauges should be calibrated to better than an ounce. Calibration is all around us.

Let's return to *La Grande Jatte*. Imagine a photographic equivalent of this scene. A good photograph contains sharp edges. Thin tree limbs, for example, should be readily visible and not fuzzy. If you saw fuzzy edges, you would return the defective camera to the store. Because the camera records the scene you saw with your eyes, you would also say the camera is a "perfect detector." (It is not,[4] but, for now, let's imagine that it is.) What, then, did Seurat's brain do to the image? Seurat chose a particular detector function, the technique of pointillism, and distorted the image. Specifically, each point of light in the photograph has been spread out in Seurat's painting. Astrophysicists would say that the photograph has a sharper point-spread function than Seurat's painting.

What is a point-spread function? It has a specific definition in optics, but let's look at some analogies. Recall the old game of telephone, in which a group of people passes a message, each person receiving the whispered message, and passing it to the next one in line. Each person believes he or she passes the message along without distortion. They do not do so because they do not remember the words clearly or they alter the meaning of the words slightly. The longer the message, the greater the distortion. By the end of the line, the message is invariably unrecognizable.

Think of the original message as an ideal source of light and the line of people as the equivalent of a point-spread function. The message started out pristine and uncorrupted. Each person, corresponding to a different part of a telescope, introduced his or her specific distortion to the message. The nature of that distortion is what scientists must understand if we are to interpret our data correctly.

As light is propagated through a telescope, each component introduces its own distortion (Fig. 3.1). The sum total of the distortions is the point-spread function.

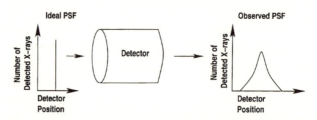

Fig. 3.1.   A representation of a point-spread function: the ideal behavior of a point source, after passing through a telescope/detector combination, is observed as a broad distribution; the width of the distribution is a measure of the distortion introduced by the telescope and detectors. The sketch of the observed point-spread function is a two-dimensional representation of a three-dimensional shape.

The clear goal of every telescope designer is the best point-spread function achievable for the funds available to purchase and grind the mirrors and build the detectors. Sometimes, a poor point-spread function is the best available. The Einstein Observatory had a point-spread function of about 1.5 minutes of arc.[5] At that time, 1.5 minutes of arc was the best available for the money appropriated by Congress. Furthermore, in 1978, only a few hundred sources were known. If no other sources were detectable, the construction of an observatory with a sharper point-spread function would have been a waste of money.

What affects the quality of the point-spread function? The quality of the mirror is absolutely critical. Here, the quality of the mirror involves more than just the materials of which the mirror is composed, although they can be important, too. A high-quality mirror requires a smooth reflecting surface. To understand this, we need to know more about X rays as photons.

Drop a stone into a still pond and ripples expand away from the point of impact. Measure the distance between the tops of two ripples. This distance is the wavelength, the physical separation between two adjacent peaks or wave crests. Measure the distance between two adjacent troughs and you'll get the same number. Peak to peak or trough to trough, the distance is the wavelength. The wavelength is a fundamental description of a wave.

The terminology we are about to learn describes light, from radio waves to gamma rays. Nearly everything we know about the universe we have learned from the light we receive from space. What most people think of as light, the astrophysicist thinks of as optical light. Let's broaden our perspective.

The separation of two ripples on the pond introduced the idea of "wavelength" but an alternative viewpoint exists. Return to the pond. Watch the waves as they reach the shore. Count the number of waves reaching the shore in a given amount of time (say, 10 seconds). Divide the counted number by the amount of time. The result is the frequency of the waves. It describes how often a wave will pass by you, the observer. Frequency is measured in cycles per second, otherwise known as hertz.[6] One cycle means two successive crests have moved past the observer in one second. One hundred hertz means that 100 waves moved past the observer in one second. (Frequency also describes sound waves: a flute most often generates high-frequency sound while a bass typically produces a lower-frequency sound.)

Frequency is consistently measured in hertz, although a prefix may be included to indicate how many hertz. Scientists report measurements of any quantity in units that are sized for convenience. For example, "mega" in front of "bucks" means a million dollars. Each particular unit applies to a limited range. We attach labels to these ranges and call them "bands": the radio band, the infrared band, the optical band, the ultraviolet band, and the X-ray band. The boundaries of the bands are, and are not, arbitrary. There are no gates on one side of which a photon is labeled "infrared" and on the other side of which it is called "optical," so in that sense, the boundaries are arbitrary. However, detection techniques for optical photons do not work in the radio band, for example. The radio antenna on a car will detect radio waves emitted by radio stations, but it will not focus optical photons. Parabolic mirrors focus optical photons, but X-ray photons pass through them.

Because of different detection techniques and the historical development of physics, our knowledge of some bands is more advanced than our knowledge of others. This is an effect of history: curiosity about the sky drove people to build larger telescopes and more-sensitive detectors to collect optical photons from ever-fainter sources. The advances in electronics during the 1920s and 1930s created the first opportunity to attempt to detect radio light from the sky; the first detections occurred in the 1930s and 1940s.[7] Because it is easier and less expensive to build telescopes on the ground, the other bands essentially had to await the development of rockets and space satellites. Consequently, we know considerably less about sources emitting in the gamma-ray, X-ray, ultraviolet, and infrared bands than we do about sources in the optical and radio bands. There is much current excitement about increasing what we know about those bands to the level of our knowledge of the optical and radio bands. Chandra, Newton, and the coming generation of infrared and gamma-ray satellites will go a long way toward filling the gaps.

Radio waves are measured in meters, but much more often, radio astronomers use megahertz and gigahertz. Megahertz means that a million waves (1,000,000, or 10^6)[8] pass the observer per second; the abbreviation for megahertz is MHz. Gigahertz means that a billion waves (1,000,000,000, or 10^9) pass by, and it is written GHz. Astronomers who work in the infrared part of the spectrum use wavelengths measured in microns. One micron is a millionth of a meter (1/1,000,000, or 10^{-6}), or one ten-thousandth (1/10,000, or 10^{-4}) of a centimeter. Optical and ultraviolet astronomers measure wavelengths in angstroms; there are 100 million angstroms (10^8) in one centimeter. Blue light has a wavelength of about 4,000 angstroms; the wavelength of red light is about 6,500 angstroms. These numbers convey sizable differences in scale, yet all of these numbers and units describe waves of electromagnetic radiation.

Can we relate these numbers to anything familiar? Yes. Look at two successive waves approaching a shore, but not so close to shore that they are starting to break. Note how far apart the crests are. The separation should be about 40 to 60 feet, or about 15 to 20 meters. Now, pick up some sand. Sift it out of your hand until you have the smallest grain you can find (about two to three millimeters). The ratio of the separation of the wave crests to the size of the grain of sand is about the same as the ratio of wavelengths of red light (~6,000 angstroms) to those of a 1-angstrom X ray. That's a crude approximation of the relationship between the wavelengths of optical light and X rays.

X rays observed by Chandra lie roughly between 1 and 120 angstroms or, equivalently, between 1.2×10^{-8} meters and 1×10^{-10} meters. Optical light falls roughly between 3,500 and 6,500 angstroms, or between 3.5×10^{-7} meters and 6.5×10^{-7} meters.

Radio waves used for television broadcasts have a wavelength of about 2 meters. The ratios cited work to within about a factor of two.

To finish the comparison, the wavelengths of radio waves that pass through Earth's atmosphere range broadly from about one centimeter to about one kilometer. Most people are between 1.5 and 2 meters tall. Radio waves of wavelength 1.5 to 2 meters carry television signals. The ratio of radio light of wavelength 2 meters to optical light of wavelength 6,000 angstroms is about three million to one. The ratio between the grain of sand and the distance to the horizon is also about three million to one.[9] Assemble the picture: the distance to the horizon is a radio wave, while the pebble is an optical wavelength; expand the wavelength of optical light to the separation between waves, and the pebble is approximately the wavelength of an X ray.

Clearly, X rays have very short wavelengths. For that reason, X-ray astronomers use a unit of energy, the kilovolt, instead of wavelength or frequency. Why measure X rays in energy units but measure optical light as a length?

A kilovolt (short for "kilo electron-volt" and abbreviated keV) is another convenient unit. The frequency of a 1-keV X ray is about 10^{18} hertz, a very large number. The wavelength, then, must be a very small number. And it is: a 1-keV X ray has a wavelength of about 12 angstroms, or 1.2×10^{-9} meters. These extremes in size make the use of wavelengths or frequencies clumsy when applied to X rays. Astronomers use energy units because a fundamental relationship exists between frequency, and energy: the energy is proportional to the frequency.[10] Wavelength and frequency are always related: if the wavelength is large, then the frequency is low. In other words, if the distance between the peaks of two ripples is large, then the complete wave will take a long time to pass by an observer, so few waves will pass the observer in a specified amount of time. If the distance between two ripples is small, many waves will pass the observer in the same amount of time.

What is the importance of wavelength and frequency? Both quantities are used to describes waves. The two quantities are complementary, so it does not matter which one is used, which means the choice can be made solely for convenience. The two quantities can be described as the equivalent of yin and yang for wave physics. If you multiply the wavelength of a wave of light by its frequency, the result is the same number regardless of the values for wavelength and frequency. That number, written "c" in physics, is the speed of light. The value of c, 29,979,245,800 centimeters per second (written $2.99792458 \times 10^{10}$ cm per sec),[11] is a constant of physics. Regardless of where you are in the universe, you'll measure the same number for the speed of light.

All the waves described above, from radio to X rays, are forms of electromagnetic radiation. We call the complete set of waves of electromagnetic radiation of all possible wavelengths the electromagnetic spectrum (Fig. 3.2 [see next page]). Electromagnetic radiation carries energy that is directly related to the frequency of the radiation. The labels attached to the wavelength bands define artificial boundaries; in reality, there is a continuum from the shortest wavelength we've

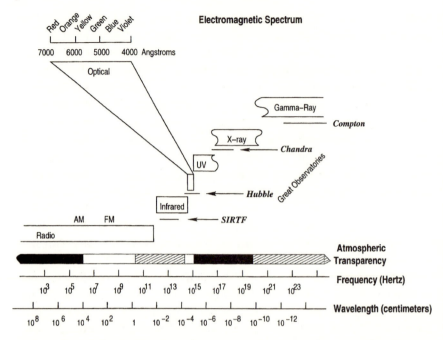

Fig. 3.2. The electromagnetic spectrum from radio to gamma rays. The horizontal axes are presented in the complementary quantities of frequency, measured in hertz or cycles per second, and wavelength, measured in centimeters. The short horizontal lines under the infrared, optical/UV (ultraviolet), X-ray, and gamma-ray bands indicate the approximate bandpasses of the four Great Observatories (SIRTF, or the Space InfraRed Telescope Facility, Hubble Space Telescope, Chandra X-ray Observatory, and Compton Gamma-ray Observatory). The bar labeled "Atmospheric Transparency" indicates the transparency of the atmosphere at each frequency or wavelength: solid black indicates a transparency of zero; diagonal hatching indicates the atmosphere is partially transparent; the white portions are completely transparent to radiation at the indicated wavelengths.

measured (about 10^{-15}) to about 10^7 meters (about 10 Hz). This is a range of 10^{22} in frequency. There is no equivalent in our experience for this range: our ears cover frequencies from about 20 hertz to about $2{\times}10^4$ hertz, a range of about 1,000. Our eyes distinguish between 1 million and about 10 million colors ($1{\times}10^6$ to about $1{\times}10^7$).[12] The national debt is about 4 trillion dollars ($4{\times}10^{12}$ dollars). The ratio of the total amount of dry land on the surface of Earth to the area of a typical beach towel is about 10^{14}.

If we think beyond the X-ray and gamma-ray domains, we may wonder whether any electromagnetic waves exist with wavelengths shorter than 10^{-10} centimeters. Recall that the shorter the wavelength, the higher the frequency and the more energy the photon carries. The more energy the photon carries, the more energetic the source must be to create the photon at all. Consequently, you'd expect very few ultra-high-energy gamma rays. Surprisingly, quite a few sources have been observed in the GeV range (giga-electron volts, or 10^9 eV), many discovered using the EGRET (Energetic Gamma-Ray Experiment Telescope) instrument on board the now-deorbited Compton Gamma-ray Observatory, the second Great Observatory. To date, most of the GeV sources are largely unknown because of the usual problem: the point-spread functions of the instruments on the Compton, while better than those on previous satellites, were very large. As a result, of the 271 sources discovered by EGRET, 170 remain unidentified.[13] The impending launch of INTEGRAL (INTErnational Gamma-Ray Astrophysics Lab), a collaboration between Europe and the United States, will have as one of its top priorities additional observations of the unidentified Compton sources, with the hope of pinning down the identification of these objects. Without a playbook, it is impossible to identify the players and understand what they are doing.

Astonishingly enough, there have been reports of TeV sources, with energies of 10^{12} eV. The emission from these objects has a wavelength of about 10^{-8} angstroms. The TeV sources detected to date fall into two categories: active galaxies, specifically blazars, and supernova remnants. Two active galaxies are known: Markarian 421 and Markarian 501. Blazars show featureless optical spectra; they dominate the extragalactic detections in the highest-energy gamma-ray band. In the context of active galaxies, astrophysicists model blazars as sources emitting a jet of material with the jet aimed directly at us. Astronomers greeted early reports of TeV emission from these objects with some skepticism. Following a coordinated observation in June 1998 between a ground-based TeV detector (HEGRA [High-Energy Gamma-Ray Astronomy] in the Canary Islands) and the Rossi X-ray Timing Explorer, during which both instruments observed a flare from Markarian 501, skepticism has vanished.[14]

On the other end of the scale are the ultralong radio waves.[15] For two reasons, we do not know whether astronomical objects emit at wavelengths longer than about ten meters. First, radio waves of that wavelength and longer are blocked by Earth's upper atmosphere. Second, the cost to build a receiver for very long radio waves is prohibitive. As a rule of thumb, one needs a radio receiver approximately the size of the waves one is interested in detecting. The largest radio telescopes are about a kilometer in diameter, a size difficult to lift above the atmosphere.

Skip a flat stone across a still pond. Watch as the stone bounces off the water. The very best skips occur with glassy, smooth water. Repeat the experiment on a windy day. Only if you're lucky will you get a stone to skip between the whitecaps. Usually, the stone will not skip at all.

The stone skipping shows the difference between a highly polished mirror and one that is not. In a radio, infrared, optical, or ultraviolet telescope, light reflects directly off a curved mirror and comes to a focus. As the energy of the photon increases, however, the light increasingly goes through a mirror placed directly in its path. How do we focus higher-energy photons then? We essentially sneak up on them: we deflect the light slightly from its straight path. By deflecting it two or

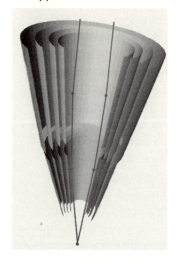

Fig. 3.3. A cutaway sketch of Chandra's mirrors showing the path an X ray takes. Note the double grazing-incidence deflections. (Image courtesy of L. Schlegel.)

more times, we bring the light to a focus. There's a simple analogy: our skipping stone. At each skip, the stone is deflected slightly off the surface of the water. If you change the stone to an X ray and the water to a mirror, the analogy is complete. The technical term that describes the deflection is grazing incidence. An X-ray mirror reflects, but only when the photon encounters it at a shallow angle (Fig. 3.3). Otherwise, the photon goes through the mirror.

The analogy also works for another reason. When you try to skip a stone off choppy, windblown water, the stone either deflects at a sharp angle to its original path or buries itself in the water. The choppy water stands in our analogy as a mirror surface that is not smooth. X rays hitting a rough surface are deflected away from a sharp focus. One of the largest costs incurred during the construction of the Hubble Space Telescope and the Chandra Observatory was that of polishing the mirrors to a high degree of smoothness (Fig. 3.4).[16] The high degree of smoothness is necessary because of the small size of X rays. To ensure a sharp focus, any roughness on the mirror surface must be smaller, by a factor of at least a few, than the size of a typical X ray. After polishing the individual mirrors, engineers and technicians assemble them with extreme care (Fig. 3.5).

Angular resolution is the ability to separate fine details. However, we do not "detect" angular resolution. Instead, we measure a position, the actual location where the X ray interacts with the detector. Position is measured in a coordinate system, say, relative to the edges of the detector. To place the photon in an image, its position must be known as precisely as possible. The precision of a position is dictated by several factors, of which two are relevant here. First, the accuracy with which the spacecraft is pointed affects the position. Each spacecraft must carry some sort of tracking system to maintain stability and ascertain the pointing

Fig. 3.4. Construction of the Chandra mirrors. Here we see one of the approximately cylindrical shells during the polishing stage. The shell shown is one of eight that make up the complete X-ray mirror for Chandra. Compare with the previous figure. (Figure used by permission of the Optical and Space Systems Division, Goodrich Corporation; the division was formerly part of Raytheon Corporation.)

Fig. 3.5. Assembly of the polished mirrors. The arclike appearance of the assembled mirrors is captured in the artist's impression of Fig. 1.3 in chapter 1. (Courtesy Kodak Corporation.)

direction. The Sun is a good target for those satellites that are not reoriented frequently. For a spacecraft studying stars, however, the exact pointing direction must be known as accurately as possible. Star trackers are usually employed for this task: star trackers are small optical telescopes that record the position of known stars in the direction of the point, stars that are easy to identify and for which accurate coordinates exist in the astronomical coordinate system. Once the data are obtained, the moment-by-moment position of the spacecraft is tracked using the known stars. The moment-by-moment track provides the information necessary to know the location in the sky from which the photon arrived. If we do not measure the spacecraft's position relative to distant stars accurately, then we will not know the photon's direction well and we will have introduced an error in the position. The moment-by-moment track is called the satellite's aspect and is defined relative to the known stars. For most satellites, the electronics record the time and aspect information; software on the ground reconstructs the moment-by-moment track.

The second factor is the point-spread function. Star trackers may reconstruct the actual pointing direction accurately, but leave the resulting image fuzzy. Compare the images of Cas A as seen from Einstein (proportional-counter data, Fig. 3.6),

Fig. 3.6. Cas A as seen by the proportional counter on Einstein. Compare this image with the next two figures. The point-spread function for the Einstein mirrors and counter had a width of about 90 seconds of arc. (Image generated from data obtained from the public archives at the High Energy Astrophysics Science Archive Research Center, NASA-Goddard Space Flight Center.)

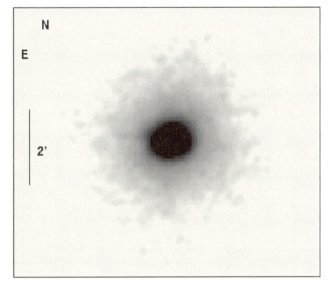

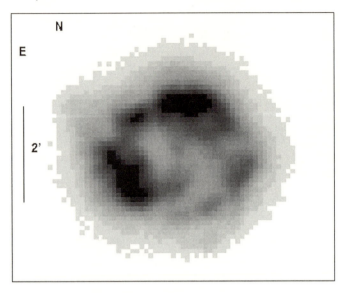

Fig. 3.7. Cas A as seen by the proportional counter on ROSAT. Compare this image with the previous and next images. The quality of the point-spread function of the detector used to obtain this data appears in Fig. 3.9. The point-spread function for the ROSAT mirrors and counter had a width of about 25 seconds of arc. (Image generated from data obtained from the public archives at the High Energy Astrophysics Science Archive Research Center, NASA-Goddard Space Flight Center.)

ROSAT (proportional-counter data, Fig. 3.7), and Chandra (Fig. 3.8). In these figures, aspect quality is not the dominant factor producing the fuzzy appearance of the ROSAT image; the point-spread function of the mirrors is the culprit.[17]

The quality of the mirrors that deflect the photons dictates the size of the point-spread function.[18] The whole point of spending so much money on Chandra is to obtain the delivery of as many X rays as possible to the smallest area. Chandra's mirrors achieve that goal, delivering more than 70 percent of the light from a

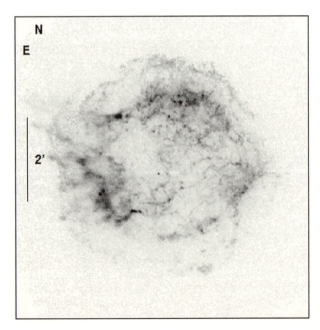

Fig. 3.8. Cas A as seen by Chandra and the Advanced CCD Imaging Spectrometer (ACIS). Compare this image with the previous two images. The quality of the point-spread function of this instrument appears in the next figure. The point-spread function of the Chandra mirrors and CCD detector has a width of about 0.8 seconds of arc. (Image generated from data obtained from the Chandra public data archive at the Smithsonian Astrophysical Observatory.)

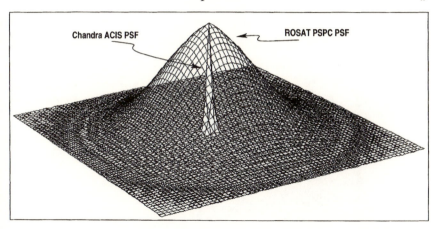

Fig. 3.9.   The point-spread function of Chandra (the narrow cone) compared to that of ROSAT (broad cone). This figure illustrates several benefits of sharp point-spread functions: (1) the signal for a source is concentrated, so locating a source is easier and can be achieved with a shorter exposure; (2) fine structure is not smeared out; (3) the amount of the X-ray background, a relatively uniform signal covering the sky, included at the position of an X-ray source is smaller. Less background leads to a higher signal-to-noise ratio. The point-spread function of the Einstein mirror plus detector is not included here, but it would be several times broader than the ROSAT point-spread function.

point source into one pixel and more than 90 percent into two pixels. Figure 3.9 offers a comparison of the ROSAT proportional-counter (PSPC) and Chandra point-spread functions.

A superb mirror will be of little use if its accompanying detector contains big pixels. Mirrors with a sharp point-spread function deliver all the photons into one pixel, eliminating angular resolution. Conversely, small pixels are not of much use if the quality of the mirror is poor. A poor mirror scatters photons away from a tight focus, distributing the events across tens or hundreds of pixels, which is a waste of money. The detector quality must match the mirror quality.

Are you thinking you have never seen big pixels? You have. Recall the last time you watched one of the television programs that send camera teams to follow the police. The United States Constitution states that you are innocent until proven guilty in a court of law. Any people arrested but not yet convicted when the show airs do not have their identities revealed because of concerns about lawsuits as well as prosecutors inveighing against media exposure and prejudicial publicity. Television stations merge the many pixels covering the face of the arrestee into one or two pixels (big pixels).

So, we have a sharp point-spread function, high-quality aspect reconstruction, and excellent positions. What have we learned? Let's explore a few results.

Elliptical galaxies are X-ray sources. Elliptical galaxies are called elliptical because they have that shape in an optical image (Fig. 3.10 [see next page]). An ellipse

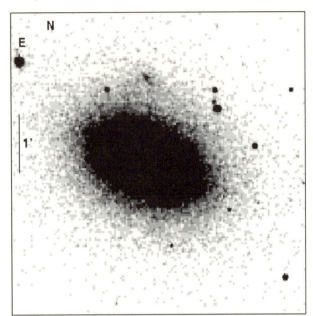

Fig. 3.10. An elliptical galaxy in the near infrared (wavelength of about 12,000 angstroms): NGC 4697. In this exposure, the center of the galaxy is essentially overexposed; the elliptical shape of the galaxy is apparent. (Image generated from data obtained from the public 2MASS data archive at the Infrared Processing and Analysis Center, California Institute of Technology.)

describes only the two-dimensional shape as seen in a photograph. The actual, three-dimensional shape is either a "prolate spheroid" (cigar shaped), an "oblate spheroid" (pancake shaped), or "triaxial" (with each axis a unique size).

Prior to the Einstein mission, ellipticals were believed to be devoid of gas and dust. Without gas and dust, a galaxy cannot make additional stars. Thus, its stellar population gradually ages. One might not predict such a galaxy to be a strong emitter of X rays. Elliptical galaxies presented a surprise to researchers when they turned out to be X-ray emitters. In retrospect, we should have realized that ellipticals would be X-ray sources because they are composed of X-ray-emitting objects: stars, particularly binary stars.[19]

Another surprise awaited us: in studying a large sample of elliptical galaxies, researchers found that some ellipticals were considerably more X-ray luminous than could be explained by adding up the X rays from individual sources. The extra emission was extended and diffuse—in other words, hot gas. Because the gas is hot, astrophysicists, when attempting to understand this behavior, naturally appeal to the other gravitationally bound objects that have hot gas: clusters of galaxies. The X-ray emission from ellipticals is composed of the emissions from hot gas and from binaries. Which dominates? What is the origin of the hot gas?

Two relationships exist, both proportional to the luminosity of the ellipticals in the optical (blue). If the galaxies are ranked by their blue luminosity, the more-luminous galaxies have more X-ray binaries and more hot gas. At first glance, this is reasonable: a galaxy is more luminous because it has more stars, so more binaries exist. In addition, a galaxy with more stars has a stronger gravitational field, so it holds on to hot gas.

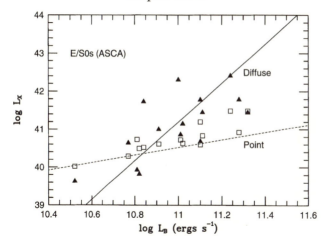

Fig. 3.11. The X-ray luminosity (vertical axis) of elliptical galaxies plotted against the blue optical luminosity (horizontal axis). The total X-ray brightness has been separated into the contribution from point sources (open squares) and the contribution from hot gas (filled triangles). The hot-gas and the point-source trends both appear to be linear, but they obey different relationships. The dashed line is a prediction of the contribution of X-ray binaries to the total point-source brightness from a galaxy. The slow increase states that more-luminous and more-massive galaxies contain more X-ray binaries. The different relationship for the diffuse emission is a topic of research. (Plot adapted from E. Schlegel et al., *Astronomical Journal* 115 [1998]: 525.)

The two relationships behave differently (Fig. 3.11). Essentially, the combined emission from X-ray binaries forms a rising floor. But above a particular value for a galaxy's optical luminosity, the X-ray emission becomes increasingly dominated by the emission from hot gas. Several questions immediately arise: do faint ellipticals have any gas? What is the nature of the hot gas? Does the X-ray-binary relationship extend across the entire range?

Observations from previous missions were insufficient to address these questions, largely because of low angular resolution. The X-ray emission from point sources blurred together and blended with the emission from the hot gas. Chandra's angular resolution eliminates these problems. Already, several elliptical galaxies have been studied, not always with the hoped-for, "clean" answer. Two faint galaxies, for which the X-ray emission should be largely the sum of point-source emissions, yield different conclusions. A Chandra study of NGC 4697[20] shows that about 75 percent of the total X-ray emission lies in the combined emissions of point sources (Fig. 3.12 [see next page]).[21] For its blue luminosity, the fraction is about right. NGC 1553, however, shows that 75 percent of its X-ray emission comes from hot gas, which is higher than expected for its blue luminosity.[22] Clearly, either a random factor is at work or we're missing something. What dictates whether hot gas or point-source emission dominates for a given optical luminosity? Right now, we do not know. One model[23] postulates that matter, lost from stars during their normal evolution, is heated by supernovae (exploding stars). The hot gas escapes the gravity well of the galaxy, eventually dissipating in the space between galaxies. This model implies dominance

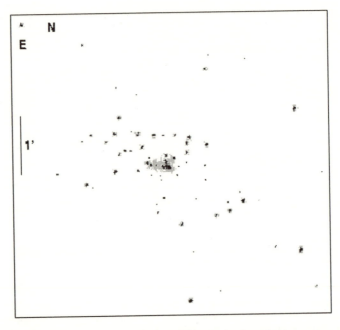

Fig. 3.12.   NGC 4697 in the X-ray band, imaged using Chandra. Several point sources are visible, as well as some diffuse emissions near the center of the galaxy. NGC 4697 lies near the low end of the L_x-L_B plot of the previous figure, so it should show little diffuse emission. For this galaxy, the Chandra image verifies the expectation. (Image generated from data obtained from the Chandra public data archive at the Smithsonian Astrophysical Observatory; used by permission of Dr. Craig Sarazin, University of Virginia. The original image appeared in *Astrophysical Journal* 556 [2001]: 533.)

by the hot gas or point sources is essentially random: one galaxy may have experienced more supernovae than another and possesses more gas. Future Chandra observations will provide the data necessary to address the question.

Next, a subject close to home and a solution to a 35-year-old mystery. Recall that the initial rocket flight in 1962 did not detect the Moon. The Moon remained undetected until 1990, when ROSAT, nearly 30 years after the initial rocket experiment, successfully found it (Fig. 3.13, p. 51).

ROSAT, which stands for Roentgensatellit, named to honor W. Roentgen, was an imaging satellite. A collaboration among Germany, the United Kingdom, and the United States, the satellite carried two proportional counters and a high-resolution imager. The mirrors and proportional counters were of the highest quality ever launched as of 1990; the mirrors delivered a view about three times as sharp as the view from the Einstein mirrors, providing a good look at the X-ray sky. Launched in June 1990 on a United States Delta rocket, ROSAT spent the first six months of its life carrying out a sensitive, all-sky survey. Following that period, observing time became available for pointed observations. The survey uncovered between 70,000 and 100,000 X-ray sources, about twice as many as had been predicted based on the small patch of sky surveyed by Einstein. The survey also mapped

Color Fig. 1. Newton's First Light image of 30 Doradus, which is a star formation and supernova remnant complex in the Large Magellanic Cloud, a satellite galaxy to our own. The colors indicate the energy of the X rays: red, 0.3-keV, to blue, 5.0-keV. The large bluish arc to the right of center is the supernova remnant 30 Doradus C; the brightest source just to the left of center is the supernova remnant N157B. The blue point sources near the image center are background active galaxies visible through this region of the Large Magellanic Cloud. (Image courtesy of the European astronomy journal *Astronomy and Astrophysics*, by permission of the journal's editor, Dr. C. Bertout. The image originally appeared in K. Dennerl et al., *Astronomy and Astrophysics* 365 [2001]: L202.)

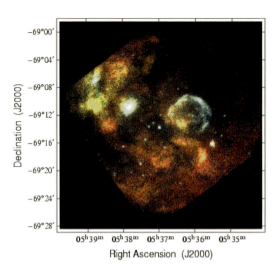

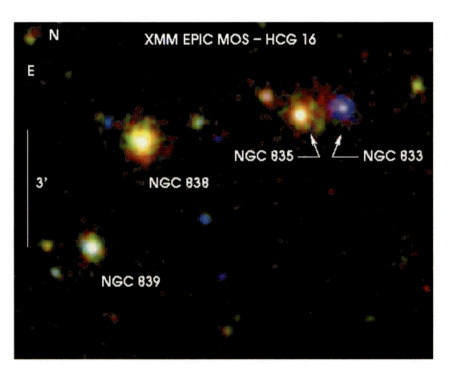

Color Fig. 2. Another First Light image from Newton: a group of galaxies known as Hickson Compact Group 16 (the 16th entry in P. Hickson's catalog of compact groups). The colors indicate the following energies: red, about 0.8-keV; green, about 1.5-keV; and blue, >3-keV. Notice that NGC 833 is very blue, while the nuclei of the other galaxies in the group are redder. (Image courtesy of the European astronomy journal *Astronomy and Astrophysics*, by permission of the journal editor, Dr. C. Bertout. The image originally appeared in M. Turner et al., *Astronomy and Astrophysics* 365 [2001]: L110.)

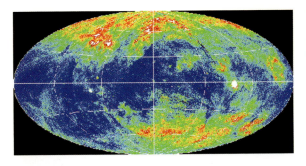

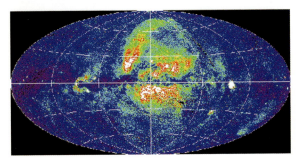

Color Fig. 3. ROSAT all-sky survey of the diffuse emission in our galaxy in two bands, the 0.25-keV band (top) and the 0.75-keV band (bottom). In both figures, all of the individual point sources have been removed. The data are plotted using galactic coordinates; the center of the galaxy lies at the center of the image, while the plane of the galaxy extends from left to right across the equator of the plot. Note that most of the emission (bright colors) is concentrated near the poles of the galaxy (top) but near the center of the image (bottom). Gas and dust in the plane of our galaxy easily absorb 0.25-keV X rays, while 0.75-keV X rays are sufficiently energetic to punch through more of the absorption. At 0.25-keV, most of the background X rays come from either hot gas near our solar system or from the universe itself; at 0.75-keV, most of the X rays come from the direction of the center of our galaxy. (Figure used by permission of M. Freyberg, Max-Planck-Institut für extraterrestriche Physik [MPE]; the original image appeared in M. Freyberg and R. Egger, *Proceedings of the Symposium "Highlights in X-ray Astronomy,"* ed. B. Aschenbach and M. Freyberg, MPE Report 272 [1999], 278–81.)

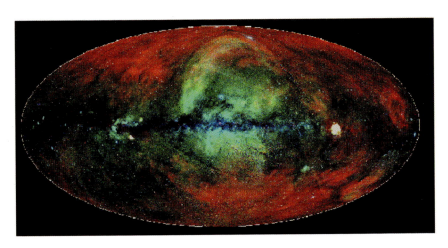

Color Fig. 4. Combined image of the three bands from the ROSAT all-sky survey of the diffuse emission in our galaxy. For this image, red indicates 0.25-keV emission, green is 0.75-keV emission, and blue represents 1.5-keV emission. Note that the center of the galaxy is essentially completely blue in color. (Image used by permission of M. Freyberg, MPE; the original image appeared in Freyberg and Egger, *"Highlights in X-Ray Astronomy."*)

Color Fig. 5. Color-coded images of Cas A as observed by Chandra. For this image, the data were filtered into energy bands, with red corresponding to energies of 0.9 to 1.05-keV (approximately the energy bounds of emission lines of neon), green for energies of 3.8 to 4.1-keV (the energy bounds of emission lines of calcium), and blue for energies of 6.5 to 6.8-keV (the energies of emission lines of iron). Note the red versus green "lagoons" on the east and west sides of the remnant and the greenish filaments in much of the interior. (Image generated from data obtained from the Chandra public data archive at the Smithsonian Astrophysical Observatory.)

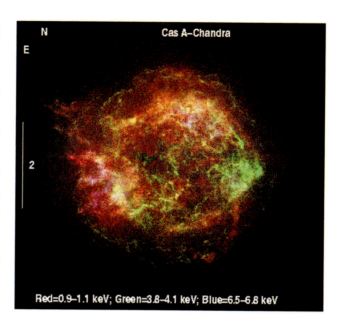

Color Fig. 6. Cas A, using the identical data from the previous image with one change: green now represents energies between 1.7 and 1.9-keV (the approximate energy bounds of emission lines of silicon). If the two figures are compared, differences in the distribution of calcium and silicon are visible. Note the green filaments on the east side. (Image generated from data obtained from the Chandra public data archive at the Smithsonian Astrophysical Observatory.)

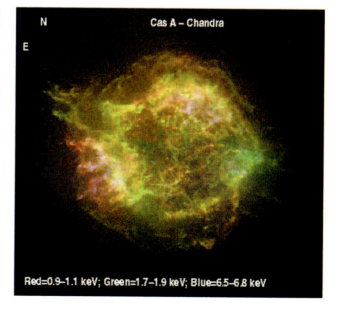

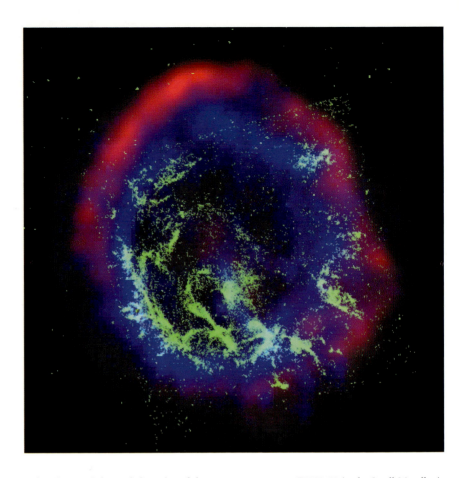

Color Fig. 7.   Color-coded overlay of the supernova remnant E0102-72 in the Small Magellanic Cloud. The colors correspond to the radio band (red), optical (green), and X ray (blue). The optical image was obtained with a narrow-band filter centered on the oxygen-emission line. Note the displacement of each color relative to the others; the X-ray emission generally lies within the radio emission, while the optical emission arises from filaments filling the center. The image of E0102-72 is a composite of three images: the Chandra X-ray image, by permission of Dr. Terry Gaetz, Smithsonian Astrophysical Observatory; the HST optical image, by permission of Dr. Jon Morse, University of Colorado; and the ATCA radio image, by permission of Dr. Shaun Amy, University of Sydney. Dr. Gaetz combined the three images into the composite that appeared in Gaetz et al. (2000) and is used by permission.

Fig. 3.13. ROSAT looks at the Moon. Note the crescent shape, which indicates the sunlit portion of the Moon, and the silhouette of the unlit portion. High-energy particles from the Sun hit the lunar soil, causing fluorescence and producing the sunlit portion. The unilluminated portion of the Moon blocks X rays from the X-ray background, producing a silhouette. (Image used by permission of the ROSAT Mission and the Max-Planck-Institut für extraterrestrische Physik.)

the diffuse X-ray emission and hot gas located in our galaxy. Color Figure 3 (see insert) shows two different bands from the all-sky survey, the 0.25- and 0.75-keV bands. (Not shown is a third band at 1.5 keV; it closely resembles the 0.75-keV band, but with a stronger concentration toward the center of the image.) In both images, researchers removed the point sources, so we see only the hot diffuse gas. Both images appear in galactic coordinates; the center of our galaxy lies in the direction of the center of the image and the galaxy spreads out left and right through the middle of the images.[24] Notice how the 0.25-keV gas seems to avoid the center and plane of the galaxy. The disk of our galaxy contains considerable gas and dust, which absorbs lower-energy X rays, explaining the lack of 0.25-keV emission along that plane. At 0.75 keV, however, the energy of the X rays is sufficient to punch through some gas and dust, so we see sources in the direction of the center of the galaxy. We cannot say whether the sources lie at the center of the galaxy or just lie between us and the center, because we need to know the distance to those sources. Note that X rays with energies near 0.75 keV are absorbed if the concentration of gas and dust is sufficiently high. Color figure 4 (see insert) shows the sum of the bands shown in color figure 3 and the 1.5-keV band.

Returning to the moon, we ask: was its detection an important observation (Fig. 3.13)? There are two answers; the first is no. The initial experiment had as a goal the study of the X-ray spectrum of the lunar soil illuminated by X rays and charged particles from the Sun. Identification of materials using X rays is carried out all the time in the laboratory, so the proposed experiment was an interesting application of similar laboratory work. However, the Apollo Moon landings provided more details about the composition of the lunar soil than would have been possible from the X-ray spectrum obtained from the rocket, especially given the state-of-the-art detectors of that day. The spectral resolution simply did not exist.

The second answer is yes: the lunar image was important. It showed us that the lunar soil does give off X rays because of the solar-charged particle flow. That emission explains why the Moon shows an illumination pattern similar to an optical image. The lunar shadow represents the nonilluminated portion of the Moon that faces away from the charged particle flow. But why is the dark side silhouetted? The dark side of the Moon must block X rays from farther away. This silhouette makes the lunar image important. What is the source of these X rays?

The silhouette exists because of the X-ray background. The background, prior to the launch of Chandra, was an unresolved source of emission first discovered during that rocket flight in 1962. Each subsequent imaging of the X-ray emission has tackled the issue of the X-ray background.

The impact of solar-charged particles on the upper atmosphere of Earth constituted one of the original explanations for the X-ray background. The detection of the shadowed X-ray background is fundamental because it demonstrates that the background lies farther away from us than the lunar orbit: a significant result, eliminating, as it does, the upper atmosphere as the primary explanation.

However, the lunar image may not be significant, because the clever astrophysicist concocts a model that locates the X-ray background in a halo around the entire solar system, or the entire galaxy, or perhaps something in between. Over the past few decades, we have learned that the Sun and its neighboring stars travel inside a large bubble of hot gas that likely formed when a star exploded.[25] The remnants of the supernova expanded to surround its neighboring stars, including the Sun. Recent work even suggests a specific cluster of young stars: the Scorpius-Centaurus cluster, which, about six to seven million years ago, would have been located in close proximity to the Sun.[26] The hot bubble eventually merges with the interstellar medium, but at a slow rate. This hot gas contributes to, but does not completely explain, the X-ray background because the gas is not sufficiently hot. The X-ray background has a hard spectrum; in other words, we detect emission not only at low X-ray energies but also at high X-ray energies. The hot bubble contributes nearly all of its emission at the lowest X-ray energies.

How do we figure out what constitutes the X-ray background? Consider a field of lights of different intensities. Initially, if you have a poor detector, you see all of the light added together. As you refine the spatial resolution of your instrument, you detect the brighter lights first. With continuing refinements, you see additional, weaker lights. What you originally thought was a glow covering the entire sky, you see now is a glow plus point sources. That is the essence of the X-ray background. Chandra aids this effort because of its high angular resolution by spotting the individual lights (Fig. 3.14). Basically, astronomers studying the X-ray background engage in an exercise in accounting. Take a household budget. Money enters a home through salaries and interest on bank accounts or investments, and exits

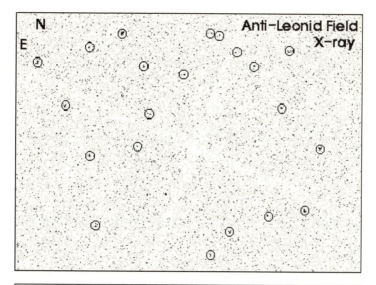

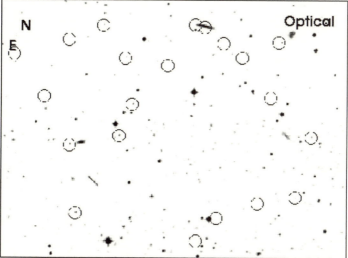

Fig. 3.14. *Top:* A long exposure from Chandra obtained during the Leonid meteor shower in November 1999. The overall science goal is the search for the contributors of the X-ray background; the practical goal is the protection of Chandra during a meteor storm predicted to be particularly rich. The circles indicate sources detected by Chandra. This image is not particularly exciting, containing as it does a few isolated point sources. The following figure provides the explanation. (Image obtained from the Chandra public data archive at the Smithsonian Astrophysical Observatory.) *Bottom:* The corresponding optical field. Now, it gets interesting. Note the following: (1) the size of the circle indicates the error on the location of the X-ray source—within the circle, with 99 percent probability, is the location of an X-ray source; (2) a few sources are positively identified with an X-ray source; (3) many of the X-ray sources have no visible optical counterpart. The small size of the detected error circle is a strength of Chandra mirrors: with ROSAT, the error circle was about 5 to 10 times as large. The larger circle increased the probability of the inclusion of additional potential optical counterparts. Consequently, a greater investment of optical telescope time was required to ascertain which optical counterpart was the X-ray source. (The background optical image was obtained from the Digitized Sky Survey available at the Space Telescope Science Institute; see the Additional Figure Credits on p. 198 that are included here by reference.)

through purchases. Imagine that you want to know where every penny goes. You write down your total income and start subtracting. You first take off the big items: mortgage, car payments, health insurance, car insurance, and the like. As you work your way down through your expenses, you likely reach an unresolved mess of small expenses: road tolls, forgotten gasoline purchases, lunch, that best-seller you bought while walking to the dentist. If you're careful, you find out what happens to each and every penny.

Studies of the X-ray background follow a similar approach. As the angular resolution increases, objects not previously resolved become so. Astrophysicists gradually eliminated the bright and moderate sources of emission, leaving an unresolved component. Among the earliest observations carried out by Chandra were long observations of apparently blank regions of the sky, or of directions along which the intervening absorption is known to be low.[27] Two of these regions correspond to the deep images obtained with Hubble: the Hubble Deep Fields (north and south, known in the professional literature as HDF-N and HDF-S). Chandra's field of view exceeds Hubble's in size, providing a larger sample of the sky (in the literature, these have become known as the Chandra Deep Fields, CDF-N and CDF-S). Newton also has examined the same regions. The value of multiple observatories looking at the same portion of the sky can not be overstated. Each provides a check on the others; in addition, the variability of the sources in the deep fields stands out. The Newton team reports that about 60 percent of the X-ray background resolves to individual sources. Three independent Chandra teams report resolving about 75 percent of the background.[28] Figure 3.14 shows a field, serendipitously observed, that was used to study the X-ray background. Predictions of the strength of the Leonid meteor shower in November 1999 ran high. As a precaution, mission controllers pointed all orbiting observatories away from the incoming meteor stream. To make use of the forced pointing, Chandra collected X rays from the "anti-Leonid" direction, shown in Figure 3.14 (p. 53).

Two key results are sought in the deep observations: the nature of the resolved sources and the behavior of the log N–log S relationship. Note that some of the detected X-ray sources in Figure 3.14 possess no optical counterpart. What are these sources? Presently, they remain a mystery. One of the teams also carried out optical identifications,[29] obtaining spectra to measure their redshifts. The harder sources on average show lower redshifts than the softer sources. The additional hardening of the spectra may result from excess obscuration as predicted by models of active galaxies. One model to explain the variety of active galaxies postulates that the viewing angle dictates the range of behaviors that we call "active galaxy" behaviors. If that model is correct, it predicts that an as yet unobserved population of active galaxies exists; these galaxies are expected to show hard spectra because the viewing angle cuts through a large quantity of absorbing gas, which eliminates the low-energy X rays. The softer objects have optical counterparts that are apparently normal galaxies, although some of the galaxies show evidence of enhanced star formation.

The behavior of the log N–log S relation depends on the distribution of sources in the universe. Consider an archery target lying flat on the ground. If you toss a number of small pebbles onto the target and subsequently count the number of pebbles per ring, you expect the number to increase. Why? The area of the outermost ring is larger than the area of the center; if the pebbles land on the surface in a random manner, then more of them will land in the outer ring and fewer in the inner ring.

Similar logic applies to X-ray sources in the universe. We divide space into concentric shells (instead of rings: three dimensions instead of two). If X-ray-emitting objects are distributed approximately uniformly throughout the universe (ignoring small-scale assemblages such as X-ray-emitting objects in galaxies), then we expect the number of sources per shell to increase as the distances get larger.

We do not know the distance to most of the objects, but we can measure the amount of energy detected from the individual sources. From this we construct the log N–log S plot. Log S is the amount of energy detected; log N represents the number of sources emitting the specified and detected amount of energy. In general, the log N–log S curve should increase toward the faint end: the number of faint sources should be much larger than the number of bright sources simply because the volume of a more distant shell is larger than the volume of a shell close to us. This approach makes a prediction: if sources are distributed uniformly, if they do not evolve, and if the universe is Euclidean, then the number per shell should increase in a known manner: as the detected energy to the 3/2 power ($S^{3/2}$). "Euclidean" means that the geometry of the universe is flat, like an archery target, rather than curved either positively (Riemannian), like the surface of a globe, or negatively (Lobachevskian), like a horse's saddle.

Astronomers search for deviations from the predicted behavior; if deviations exist, at least one of the assumptions must be incorrect. We know sources are not distributed uniformly across the sky in the immediate neighborhood. To investigate deviations from uniformity, we must search across large regions as well as reaching to ever fainter sources. Optical data suggest that only extremely deep searches provide data for investigating the geometry of the universe.[30] X-ray probes do not yet reach those levels; within the range of X-ray surveys, sources are distributed approximately uniformly across the sky. Deviations from the expected 3/2 behavior should therefore show that sources evolve. To date, the X-ray data for hard X rays, those with energies greater than about 2 keV, show no deviations from the expected Euclidean behavior. The soft-band behavior indicates the possibility of a downturn in the plot; to phrase it differently, at larger distances, soft sources appear to be increasingly in short supply. That may indicate evolution or it may be a statistical scatter. The results require confirmation from a different team and data extending to still fainter detections. Without a doubt, someone will follow up that possibility in the coming years.

ROSAT officially died on September 20, 1998, when it accidentally pointed too closely toward, or at, the Sun. More than 650 scientists used ROSAT over its more-than-eight-year life. To date, scientists have published more than three thousand articles using its data. All of the pointed observations are available in public data archives, so the impact on X-ray astrophysics will be important for many years to come.

CCDs (Charge-Coupled Devices) have become a staple of the observing astronomer's life. They were first demonstrated at ATT's Bell Labs in 1970.[31] Amateur astronomers use them as well because the price of a quality set of these chips is now relatively modest. CCDs (Fig. 3.15) are also used in video cameras (camcorders) for the personal market, although their use has faded somewhat with the introduction of CMOS imaging chips.[32] Astronomers for many years used CCDs in the optical and ultraviolet bands;[33] they are, however, new to X-ray astronomy, with their first use on the Japanese-American satellite ASCA,[34] launched in 1992. How do they work? Why have they found such significant use in astronomy?

CCDs are useful for at least two reasons. First, they are digital devices; a CCD outputs an electronic signal. Once the data are calibrated, digitally, the data can be analyzed with commercially available software or software written by the scientist. The complete process, from detecting the photon to analyzing it, is direct: the output of a CCD never leaves the electronic world.[35] This is not true of photographs that must be converted from the analog (photograph) to the digital by some type of scanner, which is an inefficient process, particularly for weak sources.

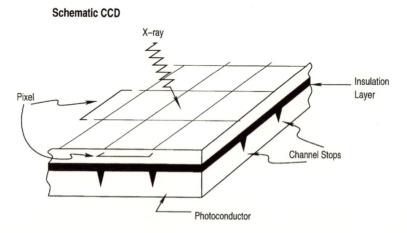

Fig. 3.15.   Sketch of a CCD, schematically illustrating the basic structure. Missing from the sketch are the electrodes that connect the pixels and by which the pixels are read. The channel stops, formed by extra insulation, define the pixel boundary.

Second, the CCD responds in a linear manner to whatever signal it is exposed to. "Linear" means that the output is directly proportional to the input. If you buy twice as much gas for your car as usual, you expect to pay twice as much. This is a linear relation. If the light that a CCD detects suddenly doubles in brightness, the output of the CCD also doubles. Linear detectors are the best because, when the output is proportional to the input, the detector becomes predictable. It can be stabilized so that one experiment can be repeated at a later date to check its results. This is crucial for scientific progress and increases the efficiency of observations. Knowing that the detector is linear, astronomers easily predict how long they must expose the CCD to light from a source to achieve a given signal, permitting efficient use of the telescope.

How does a CCD work? Imagine a set of buckets organized into rows and columns, forming an array of buckets.[36] The entire array represents the CCD and each bucket represents a single pixel. Now, use a hose to spray water across the bucket array. Some buckets will receive a large quantity of water while other buckets receive little. The water represents light, in the form of photons that a CCD collects. Measure the amount of water collected in each bucket. To do so, we must add two items to our bucket picture. First, add to the array a row, and only a row, of measuring cups. If the array is pictured as 10 rows and 10 columns of buckets, then the measuring cups form an 11th row. Finally, add to the picture a clock that can be advanced only by moving the second hand manually.

The process of measuring the water in each bucket is as follows. Advance the clock by one time step. Start with the first bucket in the row of buckets nearest the measuring cups. Use a cup to measure the amount of water in the bucket, and record the total quantity of water. When this process is finished, the bucket should be empty and the total quantity of water should be recorded. Then move to the next bucket in the same row. When the entire row has been measured, the array will consist of a row of recorded measurements, a row of measuring cups, a row of empty buckets, and the remainder of the bucket array. This is one step of the process of reading the CCD. Now, starting with the first bucket in row number 2, pour the contents of that bucket into the corresponding bucket of row number 1. Continue with all the buckets in row number 2. Advance the clock by another unit. Pour the contents of the buckets of row number 3 into the buckets of row number 2. Advance the clock by another unit. Repeat this process for each of the remaining rows of buckets. There should now be a row of empty buckets at the far end of the array. Now, return to the measuring cups and measure the amount of water in each of the buckets. When done, advance the clock one unit and pour the contents of the buckets of row number 2 into the buckets of row number 1. Advance the clock by one unit, pour the contents of the buckets in row number 3 into the buckets of row number 2, and so on. This process must be repeated until the entire array of buckets has been measured. The entire process just described is the process of reading a CCD. It is carried out electronically, but the analogy is precise.

There is only one difference between the analogy just described and a real CCD: a CCD is constructed so that the operations for an entire row (pouring all the contents of row 3 into row 2, for example) occur during one clock cycle and are carried out simultaneously. In other words, the columns are handled in parallel. How can we alter the bucket description above for the readout process to match the behavior of a CCD? Employ ten people, each with responsibility for a given column. When the clock is advanced one unit, each person will pour the contents of the bucket in the current row (for example, row 3) into the empty bucket of the previous row.[37]

ASCA was the first satellite to use CCDs for X-ray observations.[38] The spectral resolution was good but the point-spread function was quite large, on the order of two to three minutes of arc. It was a very good satellite for point sources or for large extended sources. In chapters 6 and 7 we'll see its strengths and weaknesses demonstrated.

Is there a practical benefit to the taxpayer for X-ray-sensitive CCDs and similar electronic detectors? The answer is yes.[39] There are two ways to get a precise signal: use the CCD on a bright source, or eliminate as many of the noise contributors as possible. Astronomers want to study sources in the sky that are inherently faint, so the elimination of noise contributors is a goal of their research. The study of CCDs when they are exposed to X rays leads to a better understanding of how they work, under what conditions they are best used, and the nature of the distortions they introduce into the data. CCDs, or their equivalent, then replace X-ray-sensitive photographic film. With electronic imaging devices attached to medical and dental X-ray machines, the dose of X rays can be reduced, yet the machines can still provide the doctor or dentist with the information necessary for a diagnosis. A reduced dose lowers the possibility of damage by X rays to the cells of the body. In addition, no longer must the patient wait for an X ray to be developed and read. Image-analysis software processes the image, making it immediately available.

CCDs have additional values to the astronomer: they provide spectra and they are linear devices. The next chapter discusses why their linearity is important.

4

"Sometimes you get shown the light in the strangest of places if you look at it right."

—Jerry Garcia

In the middle of winter in the Northern Hemisphere, the constellation Orion occupies a large fraction of the sky just after sunset. The three stars that make up the Hunter's Belt and the faint patch of light that defines the tip of the Sword all stand out in a dark sky. That patch of light is actually the Orion Nebula, a region where new stars, or protostars, are forming or have just recently graduated from the stellar equivalent of kindergarten. Optically, the nebula contains the Trapezium, four stars sitting in a faint cloudy patch (Fig. 4.1). The light of the stars of the Trapezium dominates the nebula. These stars are all young, hot stars, fresh from their formation. Their light illuminates the nearby gas and dust from which they were formed.

The Chandra image of this region reveals about a thousand stars (Fig. 4.2 [see next page]). One of the big surprises of the past 15 to 20 years is not only that stars emit X-rays, but also that stars undergoing formation emit X rays. Molecular clouds, from which new stars form, are vast volumes of gas and dust that exist in our galaxy. The

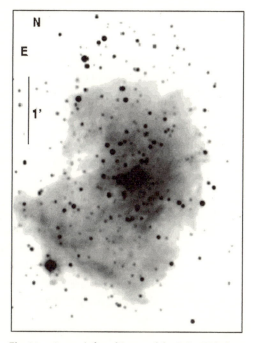

Fig. 4.1.   A near-infrared image of the Orion Nebula, a region of considerable star-formation activity. Note the profusion of point sources as well as the diffuse emission across much of the image. (Image generated from data obtained from the public 2MASS data archive at the Infrared Processing and Analysis Center, California Institute of Technology.)

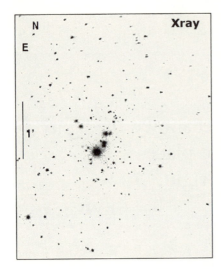
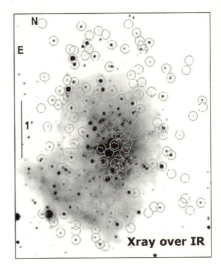

Fig. 4.2. *Left:* The Chandra observation of the Orion Nebula. Each point source is readily visible and many can be matched directly with a counterpart visible in the near-infrared image in the previous figure. *Right:* Circles indicate the X-ray source positions. (Images generated from data obtained from the Chandra data archive at the Smithsonian Astrophysical Observatory.)

gas in these clouds, with internal temperatures of about 10 degrees Kelvin, exists as molecules, atoms, and ions, depending on the sources of radiation lying nearby.

The process of star formation takes several million years. The original description belongs to Pierre-Simon Laplace in his *Exposition du Système du Monde* (1796). Laplace describes the formation of the solar system from a rotating disk of gas; this process is called self-gravitation formation. In the mid-1960s C. Hayashi summarized the research on star formation;[1] the overall picture has not changed, but many details have been added.

A disturbance, such as a shock from a nearby supernova, pushes a portion of the cloud together, raising the local density of matter and increasing the local gravitational field. Additional matter begins to fall toward that local region because of the increased mass. The infall continues, causing the center to heat up and increasing the pressure. Eventually, the temperature at the center of the dense core reaches the critical threshold and the first nuclear reactions commence. The outgoing radiation creates sufficient pressure to resist the weight of the overlying layers. What started as an amorphous cloud now has a new life as a star. This brief discussion skates over a subject that, in the past 10 to 20 years, astronomers have realized is extremely complex.

The complexities added include outflows, magnetic fields, and accretion disks. The simple picture of a collapsing cloud of gas producing a star has been supplemented by stages of development. During each stage, the protostar alternately gathers and releases energy as it gradually develops toward full-fledged stardom. The energy is supplied by gravitational contraction; the central temperature of the star is not yet sufficient to sustain nuclear reactions. The initial collapse to a protostar requires about 10,000 years. During this stage, the cloud's temperature is low (about 30 K) but slowly increasing. Protostars occupy a volume about 1,000 astronomical units (AU) in radius (or perhaps ten times as large as our solar system,[2] one astronomical unit is defined as the mean distance between Earth and the Sun, about 149 million kilometers). The collapsing protostar remains embedded within the parent molecular cloud; as a result, the developing stars appear to be detected only in the radio band. Chandra observations will establish whether protostars at this stage are X-ray sources.

A magnetic field threads the contracting protostar; as it shrinks in size, the magnetic field increasingly dominates its behavior. This stage lasts for perhaps a hundred thousand years, and by its end, the protostar's radius is a few hundred astronomical units. The energy and the flow of matter from the near-star prevent much of the surrounding gas from accreting onto the protostar. These objects emit X rays but generally remain concealed within their parent clouds. X-ray surveys with ROSAT uncovered large numbers of embedded protostars; the sharper point-spread function of Chandra has uncovered many more. Additional evolution ensues that shrinks the radius and raises the temperature at the core. Once the core temperature and density reach critical thresholds, nuclear reactions commence in the core and the object officially becomes a star. The outflow of radiation blows a bubble of space around the star; the bubble's radius is larger for increasingly massive stars. If the star was embedded near the cloud's surface, the radiation outflow forms a blister on the cloud; continued outflow disrupts the blister, exposing the star for all to see.

We glean the knowledge that protostars emit X rays by comparing the positions of stars detected in the infrared band with the Chandra images. Light at infrared wavelengths essentially passes through gas and dust clouds as if they were not present, thereby giving us a peek behind the dust curtain. We do not get that peek if we observe only in the optical band, for example. Gas and dust also block soft X rays, X rays with energies below about one kilovolt, while X rays with higher energies are not blocked.[3] The discovery that newly formed stars emit X rays immediately raises several interesting questions. Just when do protostars start to emit X rays? How do they do so? How bright are they?

Why do we want to know how bright an object is? When we say something is "bright," what do we mean? For an emitting source such as newly formed stars, we basically measure the number of photons "leaking out" to our eyes or instruments. Picture the plumbing system of a house. If there are no leaks and none of the spigots or taps is open, then no water will flow. A leak is a loss of water from what should be a closed system. You can think of radiation from an astrophysical source as a leak of energy. That leak must be supplied by something. The rate of the leak in a plumbing system can tell a plumber about the nature of the leak. A fast leak, for example, may mean a split pipe; a slow leak might mean a valve needs to be tightened. The rate at which energy leaves an astrophysical object—the luminosity of the object—provides clues to the nature of the source of that energy. Those clues provide insight into the physical processes determining the behavior of our object of study.

What we want to know is how bright any particular object really is, as if, for example, we were standing next to it (disregarding whether we can, or even want to, stand next to our object). The goal of measuring the light as if we were next to the object provides the ideal information, not diluted or blocked by intervening distance, gas, dust, or absorption of any kind. To start thinking about the energy supply requires an accurate assessment of the energy radiated to space. We cannot measure brightness; we just count the number of photons collected by our detector. Brightness is then a conversion from the number of counted photons. The brightness that we infer depends on several factors, including the collecting area of the telescope's mirrors and the efficiency of detection of the detectors.

Anyone who has ever tried to catch a nut or a small piece of candy in his or her mouth by tossing the item upward and catching it on the fall understands collecting area. Unless one has really good eye-mouth coordination, one generally misses the tossed item. Either it lands on the floor or it bounces off one's face. Why? A face has a larger collecting area than a mouth, and the floor has a still larger collecting area. If one is not so good with the toss, the game essentially becomes quasi-random. The larger collecting areas will randomly collect more.

Astrophysicists always want ever-larger collecting areas to intercept as many photons as possible. That requires larger telescopes. With more collected photons, we either observe fainter objects not previously observable, or observe bright objects more precisely. More photons are better because the greater number improves the precision of whatever quantity we are measuring. The confidence we have in any measurement depends directly on its accuracy: the more accurate it is, the more confidence we then place on the interpretation of the data. To increase our confidence, we must collect more photons. There are only three things we can do: increase the collecting area (i.e., use a larger telescope), increase the exposure time, or increase the efficiency with which we collect the photons. By how much must we increase any of these quantities?

The rule of thumb for the accuracy of a measurement is related to the square root of the number of photons we have collected divided by the number collected ($s^{1/2} / s$). If we collect 100 photons, the square root is 10, so we measure the quantity of interest to about 10 percent. Ten percent may or may not be sufficient for the science or the understanding we want to attain. If we need an accuracy of 1 percent, then we need more photons; specifically, we need to collect about 10,000 photons.

Collecting 10,000 photons may be easy if the source is bright or difficult if the source is faint. This is the fundamental reason why astronomers, regardless of the wavelength region in which they work, request bigger telescopes. None of us, whether we are radio, infrared, optical, ultraviolet, X-ray, or gamma-ray astronomers, uses "big enough" telescopes. The accuracy with which we measure a property of a source depends directly on the number of photons we can collect from that source. The number of photons collected is directly related to the collecting area. A larger telescope gives us more photons.

The collecting area is one factor; the efficiency with which our detector collects photons is another. In the laboratory, instrument builders calibrate a detector using a beam of known brightness. The ratio of the number of photons registered by the detector to the number of photons emitted by the calibration source is known as the instrument efficiency. An ideal detector has an instrument efficiency value of one: all of the emitted photons are detected. Ideal detectors do not exist, so the efficiency of a detector lies between zero and one.[4] Astronomers want detectors with the highest efficiencies possible because, for a given amount of observing time, they will collect more photons. Observing time is a limited resource, so we desire to use it as effectively as possible. If we have done the best job possible when we built the detector, we may not be able to improve its efficiency. In other words, the efficiency may be set by the physics of the detection process and not by the quality of our labor in constructing the detector.

What does all this have to do with measuring brightness? We correct the number of photons actually observed by dividing it by the efficiency. The corrected value estimates the number of photons we would have detected had we a perfect instrument, or, to phrase it differently, the actual brightness at the front end of the telescope before our instruments absorbed or scattered photons.

We have not changed our original goal; to measure brightness we need to know the rate at which our X-ray source emits photons. We now have the number of photons collected in a given amount of time across the entire collecting area of the telescope, corrected for the efficiency of detection. The rate of collection is just the number of photons divided by the length of the exposure. To measure brightness, the quantity we really need is the rate divided by the collecting area of the telescope.

The word "flux" is used for this quantity. Sunlight streaming through a window allows us to define a flux: the amount of sunlight that in a given amount of time comes through the window, with the window supplying the area. The number of raindrops falling through a hole in the roof during the night is also a flux. For photons, once we have a flux, we can convert the flux to units of energy.

But one other factor stands between us and what we want to know. A long exposure of a bright source gives us many photons; a long exposure of a faint source yields few photons. We divide the number detected by the length of the exposure to determine the rate at which we detect photons from our sources. A bright source necessarily has a higher rate than a fainter source. Does that mean the bright source is more luminous than the faint source?

Not necessarily. The brightness of an object is related to the amount of energy emitted from the object *and* the separation between the observer and the object. Picture in your mind a line of streetlights spread out over a mile. Streetlights emit essentially identical amounts of light (discounting dirty or failing bulbs), but the lights that are farther away appear fainter than the ones nearby. Look at the streetlights closely. Streetlights are generally spaced at regular intervals, so choose two, one located a quarter-mile away and a second a half-mile away.[5] Compare how bright each light appears. You will notice that the farther light is not half so bright, but appears fainter than half. Brightness does not decrease linearly as the distance increases; it decreases as the square of the distance. If an object is twice as far away, it does not appear to be half so bright, but instead one-quarter as bright.

If you cannot carry out the streetlight experiment, try the following, using a table lamp, instead. Take two sheets of white paper. Cut a perfect square out of one of the sheets (a small square, say, about the size of a postage stamp). Lay the sheet with the hole in it on top of the uncut sheet and hold both under a lamp. Look at the light falling onto the bottom sheet directly below the hole and the light falling onto the top sheet around the hole (Fig. 4.3). The light should appear to be the same brightness. Now separate the two sheets, moving the bottom sheet about twice as far from the lamp as the top sheet. Again, look at the light around the hole on the first sheet and the light reaching the second sheet. Does the light on the more distant sheet look fainter? It should. Yet, if you bring the two sheets back together, everything looks just as bright. In other words, the amount of light going through the square hole is constant, but the illuminated *area* on the more distant sheet is larger than the square hole cut into

Fig. 4.3. An illustration of the 1/r² behavior of light. The sheet of paper has a square removed from the center; the illumination of the second sheet, situated twice the distance from the light as the first sheet, covers four squares. Each square on the second sheet has one-quarter of the intensity. (Image courtesy of L. Schlegel.)

the first sheet. If you double the separation between the two sheets of paper, then the size of the illuminated square on the second sheet is four times larger. Because the amount of light is constant, its intensity must decrease. This behavior is the inverse-square law, one of the laws of physics of our universe.[6] "Inverse square" means that if the distance doubles, the effect will be one-quarter of the original. To measure the real brightness of an object (its intrinsic brightness or luminosity), we must correct for the distance between us and every object we study.

Once we have corrected our measure of the brightness of an object by taking into account the distance to the object, we then compare the luminosities of the objects we study, working toward our goal of understanding the sources of power. Using our common metric, the luminosities will run the entire gamut of energy output from low to high. Low-luminosity sources allow many explanations of physics because their power requirements are small.

High-luminosity objects, however, restrict the number of explanations, because few high-power sources exist. This is the basic implication of the discovery of quasars. Originally, quasars were dubbed quasi-stellar objects because they looked like point sources in the first photographs. Even with modern telescopes as powerful as the Hubble Space Telescope, many quasars still appear as point sources.

Astronomers realized that quasars existed billions of light-years from us.[7] When they measured the brightness (corrected for distance), they realized that the amount of energy emitted by these objects was extremely high. What sources of power could provide that amount of energy? Not nuclear power, which was often considered the ultimate source of energy even as recently as the early 1960s. Only gravitational energy, released by matter falling inward, provided the necessarily huge amounts of energy. For accretion onto a compact object, the amount of energy per gram of accreted matter is proportional to the mass of the accretor divided by its radius. If the compact object is a neutron star or black hole, then the mass is more than the Sun's mass, but the radius may be as small as or smaller than 10 kilometers. Material falling inward toward a black hole is accelerated;[8] not only does accelerating matter emit energy, but when it collides with other matter, the collision can be very energetic, thereby emitting still more energy. For the typical values just mentioned, accretion releases more than 10^{20} ergs. Nuclear energy, in contrast, surrenders at most about 10^{18} ergs (via Einstein's famous $E = mc^2$, where m represents the difference in mass of the constituents and the result of a nuclear reaction).

Astronomers believe quasars to be the active center of a "normal" galaxy; hence the generic terms "active galaxy" and "active galactic nuclei." Astrophysicists now theorize that black holes exist at the centers of all active galaxies, the proof of which is a goal of observational astrophysics.

Let's return to the Chandra image of the Orion Nebula. Look first to those unique situations where the distance to the object is relatively immaterial. The star-formation region in the Orion Nebula lies about 1,800 light-years away; the Chandra image covers a region of space about 10 light-years in diameter. This is a case where the distance may not matter: because a small region of space contains the entire star-formation region (10 light-years in size versus 1,800 light-years in distance), we can, at first, ignore the distance.

We do this solely because the cluster of newly formed stars is physically small relative to its distance. Any group or cluster of stars or galaxies can be treated in the same approximate manner. If you live in New York City, California is three thousand miles away. Does it really matter that Sacramento is one hundred miles closer to New York than is San Francisco? No, you make an approximation. For the stars within a cluster, the approximation means we may compare the stars of the cluster to one another; if we want to make a comparison between two clusters, the distances again become critically important.

Looking at the Chandra image, we see bright and faint stars, which implies that real variations exist among the stars. We've said the stars are essentially at the same distance, but we have not considered the impact of the absorption of X rays by the gas and dust surrounding the protostars. Failing to do so will affect our interpretation of any trends we find in the data between the X-ray emission and emission at another wavelength band. For example, the ratio of X-ray to infrared brightness might be a useful number: the X-ray emission tracks high-energy processes, while the infrared emission tracks low-energy processes. We need to correct the ratios by estimates of the absorption within the cloud. If we assume that the cloud size across our line of sight is also approximately the size of the cloud in our line of sight (assume the cloud is a sphere), then that assumption tells us approximately how much correction to apply to each ratio. Now we find that the ratio of X-ray emission to infrared emission is essentially constant.[9] How do we interpret that result? The astrophysical explanation is poorly understood at present; researchers search for clues among the data. Chandra and Newton will both provide necessary data: Chandra's sharp point-spread function will allow astronomers to sort the stars, matching X-ray emitters to specific optical and infrared sources; Newton's large collecting area will provide spectra of a broader range of protostars. Within the decade, we are likely to possess the answer.

At the other end of the cosmic-size ladder, consider the Newton image of the colliding galaxies in Hickson Compact Group 16 (Color Fig. 2 [see insert]). Hickson-16 is about 170 million light-years away. As an approximation, we assume that the size of the group is a small correction compared to the distance to the group (the California argument once again). Clearly, there are brighter and fainter members present in Hickson-16. Be careful, however: Hickson-16 occupies a small but non-negligible piece of the sky, so the really faint objects visible in the image may be objects more distant than 170 million light-years. These are probably background quasars or active galaxies and are unrelated to Hickson-16. The same argument applies to the Orion image, but because absorbent gas and dust permeate the region, faint objects probably are faint because they lie buried in the cloud.

The gas and dust in the cloud block the light from objects that lie behind the cloud at greater distances and leave those X rays too faint to be detected.

How certain are we that, once we count the number of X rays from a source, we know the energy emitted by the source? That depends on how well we have calibrated our instruments. Calibration is a necessary process because uncalibrated instruments are essentially worthless.

Earlier, the concept of a linear detector was raised. Linear detectors make life easier. An increase by a factor of two in the brightness of a source translates into an increase in the detector's output by a known amount. The push to use CCDs occurred for two reasons: spectral resolution and their linear nature. Photographic film is not linear. Most photographic film does not respond to any light unless it exceeds a particular brightness level (the threshold); faint objects cannot be detected at all. If the light is too bright (the saturation level), the film can become overexposed, which means that any additional light, hitting the film in the locations where it is overexposed, will not be recorded. Between these two levels, photographic film is usually linear (Fig. 4.4). Film producers work hard to ensure that their product is as linear as possible; most people never notice the nonlinearity because they photograph loved ones on bright, sunny days, or indoors using a flash, both situations falling within the linear limits of the film.

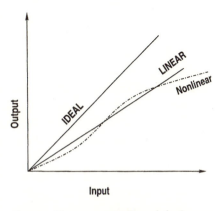

Fig. 4.4.   Linear versus nonlinear behavior, particularly for photographic film (and plates). If one doubles the input of a linear detector, the output doubles; for a nonlinear detector, that relationship does not hold, rendering the predictability of the detector problematic.

Push the film into extreme situations (too bright, too dark, or too fast) to expose the nonlinear properties of the film.

The calibration of linear detectors is decidedly easier. They are predictable, so astronomers using the detectors know to rely on estimates of the length of time needed to achieve scientific goals. Calibration of the detectors determines how they respond to X rays.

The Chandra instrument teams calibrated the detectors at the X-ray Calibration Facility (XRCF). The XRCF, located at NASA's Marshall Space Flight Center in Huntsville, Alabama, is one of a few such facilities in the world. The facility is basically a large machine used to illuminate something with an X-ray beam. The "something" can be just about any object or material inserted into the beam to gauge its response to X rays. Chandra, then called AXAF, was the first user of XRCF. The mirrors and the detectors were temporarily joined during the testing phase, and the X-ray beam served as an astrophysical source.

XRCF produces a nearly parallel X-ray beam. Parallel light is light that is not coming to a focus and is not diverging. An astrophysical source is so far away from us that the light coming from it is essentially parallel—"essentially" because the light from an astrophysical source is not parallel in the mathematical sense of the word, but the difference is much too small to measure. Why do we want parallel light? To understand how the optics will affect the incoming photons, we must illuminate the optics uniformly. Parallel light provides this illumination.

Any instrument must be calibrated to a known standard. As an example, look at a bathroom scale. How does it work? The simple versions have stiff springs under the footpads. When the user steps onto the scale, the spring compresses in direct proportion to the person's weight. The amount of compression is the measured quantity.[10] The scale is calibrated by placing a known weight on it and setting the dial so that it points to the known value. To test the scale, a different, but still known, weight is placed on it. If the scale continues to match the weights placed on it, the scale is stable.

How did instrument teams calibrate X-ray detectors for Chandra?[11] We followed a similar procedure, except that a beam of light of known properties substituted for the weights. Parallel light was essential: parallel light illuminated the mirrors uniformly and we looked for deviations from the uniform. Because the incoming light was as nearly perfect as possible, the effect of the telescope and detectors on the light was the cause of any deviations from uniform. We calibrated the mirror and detectors using a number of tests, each designed to examine a different aspect of the combination of the mirrors and the detectors.[12] Tens of people spent several months conducting the tests, producing thousands of measurements that took the calibration team many months to sift through, analyze, and interpret.

Fig. 4.5. The outside of the X-ray Calibration Facility (XRCF) at the NASA-Marshall Space Flight Center in Huntsville, Alabama. The Source Building is the small, white building at the end of the narrow pipe on the extreme left; the thin, narrow pipe is the vacuum tube. The pipe connects the Source Building with the Test Chamber that is housed inside the large building in the foreground. (Image obtained from the public image archives at NASA's Marshall Space Flight Center.)

What does the X-ray Calibration Facility look like? There is a white tube, about a half-mile long, that rests on stilts spaced about 20 feet apart over the entire half-mile length of the tube (Fig. 4.5). At one end of the tube sits the Source Building. Inside this building, the beam of X-rays is generated. The tube ends at a considerably larger building. To an observer flying over the facility, I'm sure it looks like a giant cotton swab, lying on its side on stilts and stuck into the ear of the Test Building. Inside this building, the Test Building, is a huge vacuum chamber about 100 feet long and 30 feet in diameter (Fig. 4.6). The object to be tested is rolled into the vacuum chamber (Fig. 4.7), the door is closed, the air is pumped out,[13] and the tests commence. The rest of the building contains control rooms and offices for the members of the test team. Every valve in the vacuum chamber has a gauge and a status light. Before any tests started, the facility team ascertained that the pumps and vacuum chamber were working properly.

During the calibration of the Chandra detectors, the facility operated 24 hours a day, seven days a week, for several months.[14] Calibrating Chandra was a big effort precisely because of the quality of the mirrors. Unfortunately, all too often calibration is given short attention in the rush to complete a telescope and its detectors prior to launch. Because calibration requires a completed telescope-detector combination, only a few weeks exist before an observatory is integrated into the launch vehicle. The management team for Chandra justified the expense of the calibration by the overall expense of the project. Calibration is very difficult to carry out once in orbit because so many of the targets are variable. The goal of all the tests was a calibration accuracy of about 1 percent. To date, the accu-

Fig. 4.6. A photograph of the interior of the Test Chamber during bake-out, a process that removes any impurities from the walls of the chamber. The blur in the foreground is a technician; it indicates the approximate size of the chamber. This chamber was used during the calibration of the Chandra mirrors and detectors. The mirror/detector combination was mounted on a test stand that rode the rails into the chamber; a large door sealed the near end of the chamber and the air was pumped out of it. (Image obtained from the public image archives at NASA's Marshall Space Flight Center.)

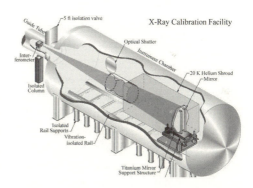

Fig. 4.7. A cutaway diagram of the interior of the Test Chamber of the Calibration Facility. (Image obtained from the public image archives at NASA's Marshall Space Flight Center.)

racy of the effective area is about 3 to 5 percent. Some previous X-ray satellites had calibration accuracies of no better than 20 to 30 percent.

To this point, we've learned how to measure an object's position and its brightness. Where does that take us? The answer depends on the object. Let's look at other X-ray-emitting objects.

Comets are essentially dirty snowballs whirling around the Sun. They are cold objects for most of their lives because they spend much of their time far from the Sun. If the comets are gravitationally bound to the Sun, then their orbit is always an ellipse. Comets not gravitationally bound travel on parabolic or hyperbolic orbits.

Two distances describe the shape of an ellipse: the peri-distance and the apo-distance, or the point of closest approach ("peri") and the point of farthest separation ("apo"). If the satellite orbits Earth, the peri-distance is called the perigee; the apo-distance, the apogee. For objects orbiting the Sun, the two names are perihelion and aphelion. For a circle, the perihelion and aphelion values are identical (by definition). For an ellipse, these values differ. The more they differ, the greater the eccentricity, the amount by which a circle must be stretched to match the ellipse.

The speed of an object moving in a perfectly circular orbit remains constant. For an ellipse, the speed of the moving object changes depending on its location in the elliptical orbit. The speed is high at perihelion and low at aphelion.[15] This behavior is one property of ellipses, and is the reason that J. Kepler used them to describe the orbit of Mars in the 1600s. Comets move in eccentric orbits—often in very eccentric orbits, with eccentricity values greater than 0.8. As a result, as a fraction of the comet's orbital period, the comet spends most of its time far from the Sun.

Now we know why comets are cold objects. For most of their lives, they are also not very interesting: lumps of frozen gas and dust moving slowly away from and then toward the Sun.[16] When a comet gets close to the Sun (say, within the orbit of Earth or of Mars), it becomes an increasingly spectacular object. The Sun warms the comet, melting its ices and releasing vapor and the trapped dust particles, both of which are pushed out behind the head of the comet, forming a long, illuminated tail. This showy display looks similar to a bridal veil trailing behind the comet's head. The veil lasts only as long as the comet is near the warmth of the Sun. Once it moves far enough from the Sun, the gases re-form into ices and the entire process starts over. It's as if an ice dancer, skating around a spotlighted point on the ice, changed her costume from a plain black body-hugging outfit to a billowing white one while near the light.

This behavior means that comets do not have sources of energy of their own; otherwise we would see them when they are far from the Sun. On this basis, we'd guess that they could not be very hot and, consequently, would not be sources of X rays. ROSAT discovered X rays from comet Hyakutake in 1996 (Fig. 4.8),[17] which came as a genuine surprise. Not only did Hyakutake emit X rays, but it was bright as well. Before we dive into the possible explanations for the X-ray emission, let us look at what we have measured.

The position of a comet changes with time—for some comets, quite rapidly.[18] Comets therefore are easily missed in a survey of X-ray-emitting sources: if a satellite looks long enough in a particular direction, it can detect rather faint objects because the photons continue to accumulate at the same location on the detector. If the object moves, however, the photons fall on a line on the detector. If the

Fig. 4.8. Comet Hyakutake as observed by ROSAT. Note that the brightness of the image appears to be symmetric around the direction toward the Sun and not about the direction of motion. This indicates that the X rays arise from a Sun-comet interaction, rather than being produced by the comet running into some matter. (Reprinted with permission of Dr. C. Lisse et al., *Science* 274 [1996]: 205. Copyright 1996 American Association for the Advancement of Science.)

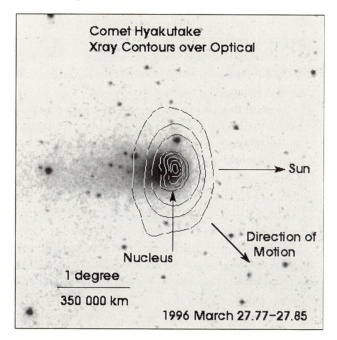

observation is interrupted for any reason, then the observer will not see the two pieces as one moving source, but rather as two separate sources. There is an implicit observational bias against discovering moving objects.[19]

To be fair, the vast majority of the objects we detect do not move appreciably within our lifetime,[20] let alone within a short observing window. Prior to the discovery of X-ray emission from Hyakutake, the only known, moving X-ray source was Jupiter.

Immediately after the discovery of Hyakutake, the ROSAT team in Germany looked closely at the data from its complete-sky X-ray survey. The team members compared the times of apparition and orbit paths of recent bright comets with the dates of observation that make up the all-sky survey ROSAT had made. When they allowed for the motion of comets through the survey observations, they discovered X-ray emission from three other comets. In astrophysics, a single detection of an unknown source of X rays is a puzzle; two such sources constitute an interesting puzzle; three make a new class worthy of detailed study. Comets therefore became the latest addition to the list of X-ray-emitting objects in our universe.

What causes a comet to emit X rays? Immediately after the discovery, astronomers proposed two explanations.[21] First, the comet, as it moves closer to the Sun, increasingly interacts with the magnetic field of the Sun. This collision of a moving object with the Sun's magnetic field could produce X rays. Second, the emission we see could really be atoms of oxygen, nitrogen, and carbon energized by the Sun's electromagnetic radiation and by the impact of charged particles ejected from the Sun. Unfortunately, the ROSAT instrument that detected Hyakutake did not have any spectral resolution. A clue to the solution of the puzzle exists in the ROSAT data, however. The X-ray emission is concentrated into a crescent on the leading side of each of the comets. That means the emission is related to the forward motion of the comet.

If space were empty, why would the forward motion of the comet make a difference? Although that question sounds reasonable, the answer is contained in the opening phrase: space is not empty. In our solar system, space is filled with the solar wind. The solar wind is a flow of charged particles from the Sun: electrons, protons, and ions. By a mechanism as yet not understood, the interaction of the charged particles with the gas and dust of the comet nucleus creates X rays. There are at least two opposing ideas about how the mechanism works. One research group believes the high-speed electrons interact with the atomic nuclei in the comet's nucleus to create X rays. This model predicts that an excess of high-energy electrons in the solar wind must exist. Such excess has been seen in comet Halley. The model predicts that the excess spectrum will be a smooth continuum. Is the excess sufficient to create X rays? A few scientists say yes, others say no.

A second model uses the ions to create X-ray emissions. When the ions interact with the gas atoms in the comet, electrons pass from the neutral gas atoms in the comet to the ions. This flow neutralizes the ions, but ionizes the comet's gas atoms. Essentially, the charge is exchanged from a set of ions to a set of neutral atoms. This model then predicts that the X rays come from the comet's ionized gas, so we expect to see a series of emission lines, in contrast to a smooth continuum in the excess-electron model. Before Chandra, none of the instruments had sufficient spectral resolution to test the prediction.

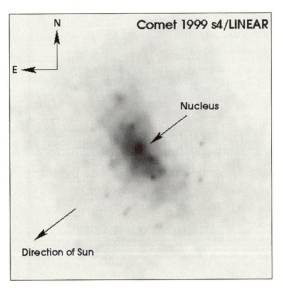

Fig. 4.9. The Chandra observation of comet LINEAR (1999/ S4) obtained on 14 July 2000 using the Chandra ACIS CCD camera. The focus of the observation was less on an improved image and more on the energies at which the comet emits X rays; the image shows X rays between 0.3- and 0.7-keV. (Image used by permission of Dr. C. Lisse, University of Maryland, and the Chandra X-ray Observatory.)

Chandra observed its first comet on July 14, 2,000 when Comet LINEAR (Comet 1999/ S4)[22] arrived in the inner solar system (Fig. 4.9). The X rays detected came from ions of nitrogen and oxygen, thereby confirming that the X rays are produced by collisions of ions in the solar wind with gas in the comet. As it happened, a solar flare had occurred on July 12. The flare increased the intensity of the solar wind. X rays from the comet varied in intensity, matching the variations in the solar wind.

Seldom in astrophysics does it happen that an unknown class of emitters is discovered, the data are analyzed, competitive theories are proposed, and additional, better observations decisively settle the question, all within the span of a few years. Research on comets is not over and done—far from it. Cometary researchers must sort out the role of magnetic fields. For example, the comets so far do not reveal the presence of compressed magnetic fields. One presumes that if the field is compressed, the higher field produces enhanced X-ray emissions. The data do not support that expectation. We have more to learn.

Instead of detecting a moving object by its emissions, we can study the long-term changes in the brightness of a source. Such studies are not unique to X-ray astrophysicists; indeed, any observer worthy of the title looks for changes because such a variation can provide clues to the nature of the object under study. The impact of comet Shoemaker-Levy 9 into Jupiter represents one of the more spectacular, once-in-a-lifetime examples. Jupiter is a known X-ray source, with the first discovery of X rays reported in 1983 by Dr. A. Metzger of the California Institute of Technology. Using the better spatial resolution of ROSAT, planetary astrophysicists such as Dr. J. H. Waite, then from the Southwest Research Institute and now at the University of Michigan, produced better maps of the X-ray-emitting locations on Jupiter. They reported X-ray emission from an equatorial belt as well as the north and south poles. The details of how Jupiter generates X rays are currently unknown.

What is known, however, is the effect that Shoemaker-Levy 9 had on Jupiter's X rays. No X rays were discovered from any of the impact points. But the emission regions at the poles brightened significantly just after the impact. The impact of the comet must have injected charged particles into the Jovian magnetic field, the particles then spiraling to the polar regions. That explanation cannot be completely correct, however, because if it were, the portion of the polar region closest to the impact point would have brightened more than polar regions distant from the impact point. That did not happen. (Another explanation ruined by data.) As a result, we do not understand in detail how the impact affected the X-ray-emitting polar regions. Perhaps observations with Chandra will provide additional clues. Or perhaps we must await either an inspired astrophysicist to provide an explanation, or another comet impact to provide more data.[23]

Supernovae are stars that explode. At first encounter, supernovae appear to be unrelated to planets and comets. Over the past few decades, we have learned that when a supernova explodes, it peppers the interstellar medium, the gas that lies between stars, with matter synthesized within the star. A star is essentially a chemical factory that uses its hydrogen and converts it to heavier elements such as carbon, oxygen, silicon, and iron. A star manufactures much of the periodic table of chemical elements heavier than hydrogen and some helium. About 10 percent of all the gold we wear, the oxygen we breathe, the iron used in tools, the silicon, aluminum, and copper used in computers, was synthesized within stars.[24]

When supernovae explode, they not only eject matter into the interstellar medium; they also emit radiation across a broad swath of the electromagnetic spectrum. Two primary mechanisms produce X rays in supernovae.[25] First, the outgoing shock wave, which has blown the star apart, compresses and heats any matter that lies in the space beyond the star. That compression and heating produces X rays. Second, the material synthesized in the explosion is usually radioactive. Radioactive materials emit gamma rays. Gamma rays degrade to X rays during collisions with matter; this process is known as Compton scattering.[26]

The first discovery of X-ray emission from supernovae occurred in 1980, making supernovae a relatively new branch of X-ray astrophysics. To date, about a dozen supernovae have been observed to emit X rays.[27] Several of them have simply been detected, so we do not know anything more about them. For the others, most of the observations have been devoted to constructing long-term light curves to see how the X-ray emission changes. Most are fading as expected; one, however, has been constant since it was found in 1993. We interpret these observations to indicate that a very dense cloud surrounds the star.

What is the scientific value of the X-ray emission from supernovae? We do not understand the details of the last years of a star's life, particularly for stars more massive than the Sun. We know that stars lose mass as they near the end of their lives, either by blowing off shells of material or by producing a steady stream of matter (a stellar wind). Either path creates a dense cloud around the aging star. The amount of material lost can be critical to the star's evolution. Massive stars are supposed to explode as supernovae and leave behind either a neutron star (which may become a pulsar)[28] or perhaps a black hole. When the supernova shock moves into the material surrounding the star, it compresses and heats it, generating radio waves and X rays as a signature of the process. All things being equal, the brighter the X-ray emission, the more dense the material. From the amount of X-ray emission, we estimate how much matter has been lost from the star. The estimate forms an important constraint for simulations of the entire life of the star; in other words, we obtain one more piece of the puzzle of a star's death. The X-ray emission may also provide observations of the chemical composition. Mass loss is a phase of stellar evolution that we do not understand well.

Finally, consider the case of LP944-20. It has an uninspiring name, yet it is a member of an important class of objects known as brown dwarfs. Brown dwarfs are failed stars, stars that do not possess enough mass to sustain significant thermonuclear reactions at their cores. Thermonuclear reactions are temperature sensitive and are initiated only above a minimum temperature of about 10 million degrees. LP944-20 has a mass about 60 times the mass of Jupiter, or about 0.06 times the mass of the Sun. It happens to be among the best-studied brown dwarfs because it lies about five parsecs, or about 16 light-years, from us in the constellation Fornax. There may be perhaps fifty brown dwarfs within 50 to 75 light-years of Earth.

Jupiter emits X rays; does LP944-20 emit X rays? A Chandra observation was scheduled to take a look. Quite serendipitously, Chandra caught LP944-20 undergoing a flare, the first flare observed from a brown dwarf at any wavelength (Fig. 4.10). What caused the flare? Currently, with one detected flare, we simply do not know. Perhaps LP944-20 undergoes flares similar to solar flares, which are magnetically driven eruptions of matter off the surface of the Sun. Magnetic fields play a significant role in the formation of stars; there is no reason currently known to expect that role to diminish after the protostar phase.

How often does LP944-20 flare? We do not know, but the existence of the flare places LP944-20 in the class of objects known as variables. Time variability is the subject of the next chapter.

Fig. 4.10. A flare on the brown dwarf LP944-20. These images may look uninspiring, but they represent the first detection of a flare from a brown dwarf at any wavelength. The top image shows the field before the flare; the bottom image shows the field during the flare. The stars in the boxes are identical in both images. The circle shows the location of the brown dwarf. (Image generated from data obtained from the Chandra public data archive at the Smithsonian Astrophysical Observatory; the original description of the observation appeared in R. Rutledge et al., *Astrophysical Journal* 533 [2000]: 141.)

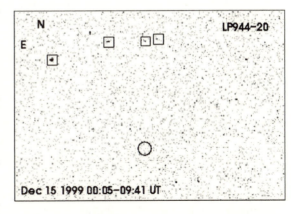

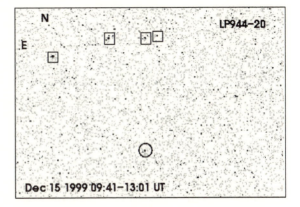

5

Veritatem dies apertit.
(Time discovers the truth.)

—Seneca (the Younger)

Raise the gun, pull the trigger, start the timer. The runners, previously tensed, accelerate out of the starting blocks toward a strip of ribbon one hundred meters away. Stop the timer as each runner crosses the line. To a scientist, this is the description of a timing experiment.

Consider the view of the timer: once started, it ticks off time until the plane of the finish line is broken by a speeding runner. In other words, the timer records the arrival of each runner as he or she breaks the line. The interaction of the runners and the finish line produces a record of the arrival times of the runners. What information may we extract from this experiment?[1] We know the distance covered by the runners and the length of time it took to cover that distance because all of the runners started at time zero. From those two pieces of information, we can calculate the average speed of the runners.[2]

How does the runner analogy compare to an astrophysical timing experiment? Instead of a finish line, we have a detector that records the time of arrival. There are, however, three fundamental differences between the runner image and the astrophysics version, all related to the information we gain from the experiment. First, we do not know that all of the photons start at time zero; in fact, we may safely assume they do not. All that we can measure is the time at which the X ray arrives. Second, we do not, in general, know the distance to the source. Third, we do know the speed of the photons: they all move at the speed of light.

Measuring the arrival time of the X ray is relatively easy and has been so ever since the first rocket flight. Each satellite contains an accurate clock. When the electronics sense that an X ray has arrived, the time from the clock is recorded. The arrival time of a photon by itself does not necessarily tell us much at all. In general, there is nothing to be learned from knowing that an X ray arrived at the satellite's detector at 11:59:16 A.M. eastern standard time on a specific date. What is important is the relationship of the arrival time of a particular photon to the arrival times of all the other photons. If we plot the number of photons versus time, for example, we might notice that at a particular time, more photons arrived than did before or after that particular instant. This type of plot is called a light curve because it describes how a specific quantity changes with time. For astrophysicists, the specific quantity is the number of photons in a particular band of the electromagnetic spectrum.

Bankers might wonder how many people use an automated teller machine (ATM) at different times of the day. If they plotted the number of people using the machine versus the time of day at which the people used the machine, they would notice that many people arrived during lunch hour and near dinnertime, but essentially no one at midnight. Why is that? The bankers realize that users of ATMs are grabbing some money so they can buy lunch or before going out on the town for the evening. Such information is valuable to them in choosing when and how often to replenish the machine's supply of money. In other words, from a time history, patterns of behavior are noted and subsequent actions adopted. Astrophysicists study a light curve, a time history of photons arriving from a particular source, to infer clues about the behavior of the emitting source.

At this point, we have a plot of the number of X rays versus their time of arrival. Astrophysicists, however, cannot just leap to any conclusion. First, are the excess photons statistical fluctuations? Contrast the number of bills ejected from an ATM with an astrophysical emission process: no guarantee exists that a fixed number of photons arrives at our instrument each second. The process may instead be steady in an average sense. During the past hour, for example, the number of photons of sunlight passing through my window has been on average the same number because over short periods of time, the Sun is a steady source.[3] But if I count the number of photons that arrive this second and compare it to the number that arrive the next second, I may get different numbers even if there are no clouds in the sky. The variation is a statistical fluctuation.

You might think that this cannot be worth pursuing: how can we untangle the statistical fluctuations of a steady source from the changes that a variable source shows? In the previous chapter, we considered the differences in measuring a quantity to 10 percent accuracy versus 1 percent accuracy. Here is a case where the degree of accuracy may depend on the science goals. The astrophysicist may need 1 percent accuracy to decide whether the source is variable. With 1 percent accuracy, a source that really changes its brightness might be easily detected, whereas at 10 percent accuracy, the changes in brightness could remain masked. Statistical accuracy is critical in timing experiments because no additional information is available, other than the light curve, to provide a clue to the nature of the source. Let's imagine, then, that we have a statistically accurate light curve. What may we learn about an astrophysical source from changes in its brightness?

Picture a lighthouse, or a searchlight at a movie premiere. As the searchlight rotates, the beam moves around the sky (Fig. 5.1 [see next page]). If you counted the number of photons arriving at your eye each second, you would notice that there were a few seconds when the number of photons was very high. Those moments

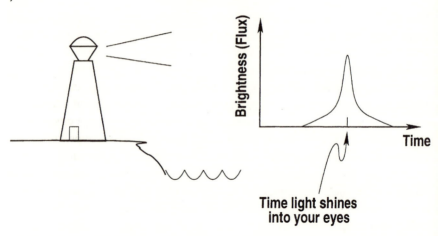

Fig. 5.1.   A light curve of a lighthouse, illustrating a familiar behavior applicable to binary stars. The figure illustrates the case of a binary star emitting light into a beam such as a pulsar, but the concept of repetitive behavior applies to normal stars. For example, in an eclipsing binary system, one of the stars partially or completely blocks the light from its companion from our line of sight for at least a short time. This example is essentially an inverse of the figure: instead of a brief flash of light as the beam crosses our line of sight; the light from the eclipsing binary is approximately constant except for a brief period when the light diminishes.

correspond to the time when the beam pointed directly toward you. At other times, the number of photons would be very low or perhaps even zero. This behavior would repeat. If you counted the number of seconds between the times when the beam was pointed directly at you, you would measure the time over which the searchlight rotates through 360 degrees. That time is the rotation period for the searchlight. If you measure the rotation period many times over the course of a few hours, you will notice that the number does not change, because the motor that drives the searchlight does so at a fixed rate.

Now point your telescope at your favorite astrophysical object. If you see similar behavior, in other words, behavior that repeats after a constant interval of time, you will undoubtedly have discovered a rotating object.[4] The time between pulses directly measures the period of rotation of the source. That number can be interesting. If it is very low—in other words, if it takes very little time for the source to rotate once—then the source cannot, in general, be very large.

How is that inference obtained? This is the light-travel-time argument we encountered earlier. Matter has a limited ability to resist being torn apart. A strong wind, for example, can break a tree branch or rip shingles off the roof of a house or break the house apart, while a mere breeze will not even lift a shingle's corner. A rotating object deforms as it spins. Place a water-filled balloon on the end of a string and whirl it around your head. If you look closely, you will notice that the balloon is not round, but flattened, and that the degree of flattening is related to the speed at which you whirl the balloon. Large astrophysical sources that are rotating quickly will flatten; if they are rotating too quickly, they will break apart as the gravitational forces that hold each half together are exceeded. Therefore, if we measure a short rotation period, we infer that the source is relatively small.

What else can we learn from looking at the light curve? In addition to rotating, some stars pulsate, or "breathe." A pulsation may first suggest a rotating object, but sufficient data will eventually demonstrate a different conclusion: not rotation but pulsation. Some patterns indicate not rotation, but revolution: one object orbits another. Generally, rotation periods are much shorter than orbital periods; Earth rotates once in 24 hours, but its trip around the Sun takes about 365 times longer.

Some sources erupt: they increase their brightness by a large amount in a short interval of time. If the increase in brightness is substantial and the length of time is short, the outburst is more accurately described as an explosion. Such stars are called novae or supernovae, depending on the size of the outburst. Finally, some variations in the light curve of an astrophysical source are measurable but do not repeat. These sources can be difficult to understand because scientists do not have so much information to study as they do when an object is repetitively variable. They know only that the object is variable. That is hardly a stunning conclusion.

Do not think that the analysis of light curves is as easy as the above paragraphs imply; timing analysis is difficult, particularly if the observatory is not in an optimal location. For timing analysis, "optimal" does not apply to the vast majority of observatories, both on the ground and in space. Why?

Timing analysis is relatively easy for the following conditions: lots of photons and a long sequence of uninterrupted observations. The "lots of photons" makes sense: the statistics will be good, particularly if the scientist searches for possible variations of the source. Why the need for the uninterrupted series?

Recall our goal: we do not know that an object is variable, so we must search for evidence of variability. If we only need to know whether an object varies, that is a trivial measurement: divide the data into subsets, compute the average for each subset, and compare the averages. The existence of variability is not necessarily useful information, but repeated variability is. The goal must be revised: we must search for evidence of periodicity. We do not know the period, so we must search all possible periods.

How to do this? The sassy answer: hand the data to a computer programmed to carry out a search for periods. That is the answer, but let's look briefly at how the computer program is constructed. Because we believe the object may be rotating, orbiting, or pulsating, we assume that any variation is gradual, where "gradual" has the usual meaning: a relatively slow expansion followed by a contraction, in the case of a pulsating star, or a relatively slow change in the perspective of an emitting object (such as the lighthouse). A gradual change can be described as a wave, where the trough of the wave corresponds to a minimum of the repetitive behavior and the crest of the wave to the maximum. We program the computer to choose a specific period and then compare the data with a wave of the chosen period. A match produces a possible period; for a mismatch, the computer tries another period. The result is a list of possible periods detectable in the data.

One of the possible periods will almost certainly correspond to the orbital period of the telescope. Why? The data consist of arrival times of photons, regardless of the bandpass. Many satellites orbit at a relatively low altitude of about 580 kilometers;[5] their orbital periods around the earth are about 90 to 100 minutes.[6] Consider ground-based optical telescopes as satellites in an extremely low orbit with an altitude of about one to two miles and an orbital period of about 24 hours. Photons can be recorded only when the object lies in the field of view of the telescope. For ground-based observatories, the object is blocked from view for about 16 hours per day. Low-Earth-orbiting satellites can view an object for about 40 percent of an orbit, or about 40 minutes. The periodic lack of photons is itself a signal impressed on the data, and that signal will be found by the period-searching algorithms.

To understand the difficulty of analyzing interrupted data, consider a stock market index, such as the Dow-Jones Industrial Average (DJIA). The DJIA is reported every night on the news, so investors can track its behavior and attempt to understand the impact of other economic factors on the index. But there are no interruptions in the DJIA. The market reopens the next business day at the same value as it had when the market last closed; the DJIA may not remain at the opening value for long, but that is a different consideration.[7] Consider, then, the difficulties of market analysis if gaps existed in the reporting of the DJIA. Let's say that every day at 12:35 P.M. eastern standard time, the DJIA would cease broadcasting its values. Imagine, too, that information from the other markets was not available, yet trading continued at the world's exchanges. The forced interruptions would skew investors' trading habits. No doubt, someone would develop a technical index to forecast a trend.

In the 1980s, the Europeans launched a satellite, EXOSAT (European X-ray Observatory Satellite), into a long, elliptical orbit.[8] All previous satellites in X-ray astronomy had orbited in low-Earth, and relatively circular, orbits. Consequently, observations obtained with them had significant interruptions. EXOSAT's long orbit enabled the first good studies of the inherent variability of X-ray sources, particularly those in our galaxy. We'll soon see the benefits.

You may still be tempted to wonder why the studies of the arrival times of X rays could be so important. During the 1960s when astrophysicists were first exploring the X-ray sky, no one understood any of the details about the types of objects that would emit X rays. The early X-ray observations, crude as they were by modern standards, gave astrophysicists sufficient glimpses that they established a fundamental insight: many of the X-ray sources in our galaxy were binary stars.

The stars look quite constant in the sky, night after night, month after month. Reassuringly constant, perhaps, or boringly constant, depending on one's point of view. Is there really anything astrophysical to learn from variability?

If you could see the X-ray sky directly, you'd have a different impression. The sky would not be reassuringly constant; instead, it is restlessly variable. X rays come from the extremes of physics: very hot gas, very strong magnetic fields, very high matter densities. A "hiccup" at these extremes produces substantial quantities of X rays. In the optical band, events generally occur slowly. At X-ray energies, events almost never take place slowly, unless the object is physically extremely large, such as a supernova remnant or a cluster of galaxies. Let's look at some examples.

Pulsars are rotating neutron stars. They are truly born in fire. A star is a relentless shoving match between the force of gravity, which pulls the star inward, and the pressure of intense radiation and hot gas, which pushes the star outward. Nuclear reactions convert hydrogen to heavier elements that slowly accumulate in the innermost portions of the core. In a star's later years, the nuclear reactions increasingly occur in the outermost portions of the core. As the core gradually runs low on fuel, the outward flow of energy diminishes, reducing the pressure that holds up the outer layers. Once a star reaches this condition, its remaining life is short. As the core exhausts its sources of fuel, it starts to collapse inward under its own weight, heating the material catastrophically. The sudden rise in energy flowing outward blows the star apart.[9] Only the collapsed core remains, forming a neutron star, a hot sphere about 10 to 15 kilometers (6 to 10 miles) in diameter. A neutron star has more mass than the Sun, but a diameter about a million times smaller. It is extremely dense: about 10^{14} grams per cubic centimeter. A teaspoon holds about five cubic centimeters.[10] Among the heaviest naturally and easily found items you can place into a teaspoon here on Earth are pieces of gold or platinum, which have densities of about 20 grams per cubic centimeter. No contest, game over, neutron star wins. Physicists create glimpses of neutron-star densities in the laboratory via collisions of heavy ions, which reach densities of about 2×10^{14} grams per cubic centimeter. But those densities last for less than 10^{-20} seconds.[11] Neutron stars will still be around when the entire universe has run out of gas.[12]

We know pulsars must cool relatively quickly because if they did not, tens of thousands of them would be visible in the sky, each glowing with a temperature of about 10 million degrees. Instead, X-ray surveys show only a few hundred. That does not mean pulsars cool off completely. Just as a hot stove takes time to become completely cool, neutron stars also cool over long periods of time. The initial drop in temperature from 10 million degrees must occur relatively quickly, however.[13] Off by themselves, isolated pulsars simply cool off. Astrophysicists are interested in isolated neutron stars because they may provide clues to the nature of the equation of state for matter at densities of 10^{14} grams per cubic centimeter. As noted above, these densities are difficult to create in the laboratory for any sustained length of time. They are small and, if isolated, they do not accrete much matter to their hot surfaces, so they are faint and difficult to find without high spatial resolution.

To find such an object an observer needs a precise position, because the number of stars increases astronomically at fainter magnitudes. In the optical band, for example, astronomers detected one isolated neutron star at magnitude 25.7, about 80 million times as faint as the eye can see; it may be the closest isolated neutron star, with an estimated distance of about 60 parsecs.[14] The neutron star, first located from its X-ray emission with ROSAT, has an estimated radius of 14 kilometers (about nine miles); that size makes locating the needle in the haystack nearly trivial in comparison. This particular neutron star happens to be passing a nearby molecular or dark cloud, so it accretes some matter to its surface. If a hot neutron star accretes matter, it becomes easily visible. Place a neutron star as one star of a binary system and you will see all manner of exotic phenomena.

I will try to build a picture of binary star physics that is better described mathematically. Gravity acts radially; in other words, regardless of the location of the central mass, the force experienced by a small quantity of matter is a direct line between it and the center, if no other forces act. Picture a sphere surrounding a central object; the sphere represents the surface of equal gravitational acceleration toward the central object. Equivalently, the sphere represents the volume of space under the direct gravitational influence of the star.[15] Introduce a second star into the picture and the same statement holds for the combined gravity field around both stars, provided you are sufficiently far from them that the two stars appear to act as one. If the two stars are in turn well separated from each other, then their individual surfaces of constant acceleration will be nearly spherical. If you are very close to one of the stars, that nearby star dictates your motion, as if the more distant star were not present. Think of the innermost moons of Jupiter: to a good approximation, the moons orbit Jupiter as if the Sun did not exist.

If, however, the two stars are relatively close to each other, the physics becomes interesting. Let's assume the two stars are fixed in space and not orbiting each other; this is unrealistic, but the limitation serves to describe the picture, and we will remove it in a moment. As you move closer to the two stars, you will feel an acceleration that is the combination of the acceleration toward the first star and the acceleration toward the second star. If you mapped out the accelerations you feel and sketched a surface of constant acceleration, your drawing would resemble a fat peanut.

Start the orbital motion of the two stars about each other. The motion adds centrifugal acceleration to the picture. Centrifugal acceleration is the outward "force" you feel as you drive too fast around a corner, or take some amusement park rides. The result of the added acceleration is a pinching of the "fat peanut" shape in the region between the two stars. Instead of a fat peanut, substitute a more typical peanut, one with two distinct lobes and a narrow connection. We are not finished yet because this is only the constant-acceleration shape you would map while at a large distance from the binary.

Move closer still toward the two stars and you reach a surface where the gravity field resembles two tears kissing.[16] This shape is called the Roche critical surface,[17] or the Roche lobe; the point of contact always lies on a line between the two stars. The lobe essentially separates the combined gravity and centrifugal accelerations of the two stars from the gravity and centrifugal accelerations of the stars individually. Inside the critical lobe, the combined centrifugal plus gravity accelerations of one star dictate all motion within that lobe; outside of critical surface, all motion is dictated by the combined gravitational and centrifugal accelerations of both stars. Again, if you're very distant from the two stars, the Roche lobe is nearly spherical, regardless of how close or far apart the individual stars are.

Earth and the Moon form a "binary planet" because the moon is a large fraction of the size of Earth. A Roche critical surface exists around Earth and the Moon, but it is not particularly interesting because Earth and the Moon are considerably smaller than the critical surface. If the Moon approached Earth closely, the Roche surface would become more important. For example, if the Moon were about nine thousand kilometers (about 5,500 miles) from the surface of Earth, the gravitational force of Earth on the near and far sides of the Moon would exceed the gravitational force holding the two sides of the Moon together.[18] Once this occurred, the Moon would split apart, perhaps eventually creating a disk of dust much like the rings of Saturn. (The rings of Saturn lie within Saturn's critical Roche limit.)

For binary stars, however, the physics can be considerably more interesting. A star can fill its Roche lobe because a star is a plasma (a fluid of charged particles). If the star expands beyond the lobe, the overflow matter is no longer gravitationally bound to the star but is attracted to its companion instead. A crude description states that matter flows through the point of contact to the companion.[19]

The definition of a binary-star system means that two stars orbit each other.[20] Matter spinning about an axis or orbiting another body possesses "angular momentum," a quantity physicists track. Because the two stars orbit each other, matter leaving either star also possesses angular momentum. The classic example of angular momentum is the spinning ice skater. As the skater pulls his or her arms inward toward the body, the skater's rate of spin increases. Unless forcibly altered, the amount of angular momentum is fixed or conserved, which is another law of physics. If skaters move their arms outward, their spin slows. Matter pushed out of the Roche lobe of one star of a binary obeys the law of conservation of angular momentum. It initially moves toward the companion star, but because of its angular momentum, it cannot fall directly to it on a radial path. Instead, it orbits the companion star. As additional material flows from the donor star, the matter builds a dense ring that gradually spreads inward and outward to form a disk of matter.

Astrophysicists call such an arrangement of matter an accretion disk (Fig. 5.2 [see next page]). As the matter jostles around in the disk, some of it loses energy and falls onto the receiving star. Usually, the receiving star is compact, the remains

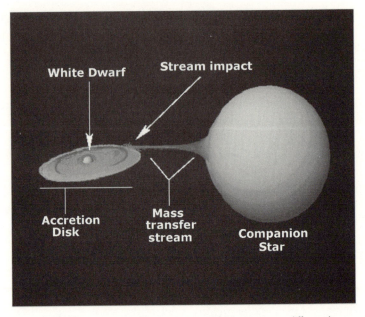

Fig. 5.2.   A sketch of an accretion disk in an interacting binary system. (Illustration courtesy of L. Schlegel.)

Fig. 5.3.   Chandra's namesake, the Indian American astrophysicist and Nobel laureate Subrahmanyan Chandrasekhar. Chandra received a Nobel Prize in physics in 1983 for research into the structure of white dwarfs. (Photo by K. G. Komsekhar, courtesy of the American Institute of Physics, Emilio Segrè Visual Archives.)

of a star that exhausted its fuel. Three endpoints await the fuel-exhausted core of a star; those ends all depend on the amount of material (its mass) within the core. If the core has less than about 1.5 times the mass of the Sun, the object becomes a white dwarf. The value of 1.5 is called the Chandrasekhar mass limit after the astrophysicist who discovered it, Subrahmanyan Chandrasekhar (Fig. 5.3).[21] If the core mass is greater than about 1.5, but less than about 3 times the Sun's mass, the object becomes a neutron star. Finally, if the core mass is greater than about 3, the object becomes a black hole.[22] Isolated stars can also create compact objects, in which case the compact star becomes our "hot but cooling" friend described earlier. In a binary system, however, "hot but cooling" translates into "very interesting."

When matter falls onto the compact star, it becomes hot quickly. If only a little matter has fallen onto the surface, then it just sits there while additional matter accumulates. Throughout the accretion process, however, the pressure at the base of the accumulating layer gradually rises. The steady accumulation of matter at the top eventually pushes the pressure at the base above a critical threshold. Hydrogen makes up most of the matter from which stars are formed. Hydrogen at high temperatures and high pressures, confined to a small volume, suffers "irreversible positive feedback." On Earth, we call such a situation a "hydrogen bomb." For the binary star, astrophysicists use the term "outburst." The size of the outburst depends on how much matter has accumulated and the nature of the compact object.

If the compact object is a white dwarf and a considerable quantity of matter has accumulated, the outburst is called a "nova" (Latin for "new star"). If the compact object is a neutron star, the bursts become more energetic because neutron stars are more massive (about two to three times the mass of a white dwarf). If the object is a black hole, all hell breaks loose. The masses of collapsing cores that exceed the upper limit for neutron stars should become black holes; black holes evolved from stars should have a minimum value, or a floor, but there is no known maximum value.[23] This means that the typical burst in a binary system with a black hole can be extremely energetic. The brightest X-ray source ever detected is a candidate black-hole binary system.[24] In 1978, an extremely bright X-ray source flared in the constellation Monoceros, increasing its X-ray brightness by a factor of about 10 million in one week. It faded completely about two hundred days later. Observations after the burst by Dr. Jeff McClintock of the Smithsonian Astrophysical Observatory to identify the source of the burst revealed an unremarkable red star orbiting an "undetected" object. The object is known to be present through the orbital motion of its companion; this motion was determined by additional work over the past two decades. The red star orbits an unseen object that has a mass greater than seven times that of the Sun.[25] Estimating the mass of the two stars in a binary system depends on the angle at which we view the binary. If the binary is viewed edge-on, as if we were looking at the rim of a plate at eye level, then the mass estimates are relatively precise. If the binary is viewed face-on (looking down on the full area of the plate), the mass estimates are poor. The X-ray nova GRO J1655-40 currently holds the record for most-precise estimates of the masses of its stellar components:[26] the unseen star has a mass of 7.02 solar masses with an error of 0.22 solar masses.[27] This object forms the best case for claiming the discovery of stellar black holes.

Let's look at what we've learned here. People observed novae for hundreds of years, even if they did not understand what they observed. Observations of novae exist in the historical records of emperors from the Korean and Chinese dynasties, particularly after A.D. 800.[28] Novae become brightest in the optical and near-ultraviolet bands. But novae occur in binary stars with a white dwarf as the compact object, and white dwarfs are the least massive compact object. Take out the white dwarf and replace it with a black hole. Leave everything else the same. Outbursts again occur, but now they are extremely energetic sources of X rays, the X-ray novae. On average over the past ten years (when satellite coverage of the X-ray sky improved), about two X-ray novae have been detected somewhere in our galaxy every year.[29]

Accretion disks themselves undergo bursts. Dwarf novae, binary systems in which the compact object is a white dwarf, undergo repeated outbursts. Bursts occur in these systems not because of a pressure-temperature relationship leading to a detonation, but because a cyclical, nonlinear relationship exists between the disk's surface density and the matter temperature in the disk;[30] this relationship essentially flips the disk from a hot to a cold state and vice versa. The hydrogen in the disk causes the flip. Neutral hydrogen does not absorb much of the outgoing radiation from the disk; if it is ionized, however, the absorption increases. The rate of matter flowing into the disk from the companion exceeds the rate draining from the disk. As material flows in, the disk's temperature increases. Near a specific temperature, the hydrogen in the disk ionizes; this traps radiation that normally escapes the disk. The trapped radiation increases the temperature still further. One might think this is a runaway situation, but it is not. As the temperature increases, the matter in the disk starts accreting onto the white dwarf at a greater rate.[31] Eventually, the rate of accretion onto the white dwarf exceeds the mass-transfer rate from the companion; in other words, the disk drains faster than it fills. The disk temperature then decreases, the hydrogen again becomes neutral, and the cycle starts over.

Amateur astronomers have followed some dwarf novae for more than a hundred years. Figure 5.4 shows the outbursts of one such object, SS Cygni (*Cygni* is the Latin

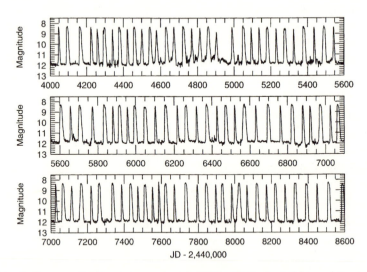

Fig. 5.4.   Twelve years of outbursts (approximately 1979 to 1991) extracted from a data set consisting of more than one hundred years of optical outbursts of the cataclysmic variable SS Cygni. Amateur astronomers have produced essentially all of the data points that define these outbursts. The horizontal axis is defined in Julian Days, the sequential count of the number of days from noon on January 1, 4713 B.C. Note the alternation of a wide burst and a narrow burst, such as the series that runs from about 6150.0 to 6800.0. The alternating bursts are often interrupted by a second narrow burst. (Image used by permission of Dr. J. Mattei, Director of the American Association of Variable Star Observers.)

form of Cygnus, the constellation of the Swan),[32] as recorded over a 12-year period by amateurs from the American Association of Variable Star Observers. Notice how regular the bursts are. Furthermore, examine the width of the bursts; notice that a wide burst is usually followed by a narrow burst. This pattern does not faithfully repeat, because occasionally two narrow bursts follow a wide burst or two wide bursts follow a narrow burst. How often? There are 93 bursts plotted; 66 of them fall into the wide-narrow pattern; there are 5 pairs of wide-wide and 5 pairs of narrow-narrow. What causes the outbursts to be regular, yet not so precisely regular as to be predictable? Astrophysicists are still working on that one.

The general name for X-ray sources in a binary-star system is "X-ray binary." X-ray binaries exhibit a wide range of behavior. Figure 5.5 presents a sample of outbursts of X-ray binaries; the graphs plot count rate (vertical axis) versus time (horizontal axis). Bursts come in lots of shapes, from the "all at once" burst of Aquilae X-1 or 0115+634 to the sputtering burst of GRS 1915+105 or 3A1942+274. All these bursts provide clues to the behavior of matter at high densities and temperatures. We do not necessarily understand what the clues tell us, but clues they are. For example, Aquilae X-1 is a low-mass X-ray binary in which the companion is a low-mass star; 0115+634, however, is a high-mass X-ray binary which means the companion is a massive O or B star. The difference sounds superficial, but the process by which the compact object is fed differs in the two types of X-ray binary. The high-mass star generates a strong wind through which the compact object trav-

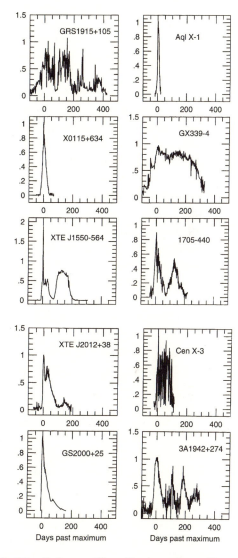

Fig. 5.5. Outbursts of X-ray binaries as observed by the various X-ray satellites. This plot shows detected X rays versus time during the outburst. It illustrates a portion of the range of behaviors observed by astrophysicists studying explosive events in these objects. (Plot generated from data obtained from the public archives at the High Energy Astrophysics Science Archive Research Center; data originally prepared by Dr. W. Chen, Dr. C. Shrader, and Dr. M. Livio, and published in *Astrophysical Journal* 491 [1997]: 312.)

els during at least a portion of its orbit. Accretion of wind material occurs only at those times. Low-mass stars cannot generate a strong stellar wind and so must feed the compact object by mass transfer through Roche-lobe overflow. Another difference exists: the nature of the compact object. Whereas Aquilae X-1 is a low-mass X-ray binary with a neutron star for the compact object, astronomers believe GRS 1915+105 to be not only a low-mass X-ray binary, but also a candidate black-hole binary.[33]

The data illustrating the burst of Aql X-1 or XTE J1550-564 or GX 339-4, all shown in Figure 5.5, were collected by the All-Sky Monitor (ASM) detector on board Rossi X-ray Timing Explorer (RXTE),[34] which was launched in December 1995. The ASM, designed and built by a team at the Massachusetts Institute of Technology, is one of three X-ray detectors on RXTE: the ASM, which sweeps across the sky about once every 95 minutes, a huge array of proportional counters, and a detector sensitive to very high energy X rays (20 to 150 keV). In addition, RXTE carries an array of "fast" arithmetic processor chips.[35] These chips can be programmed from the ground to search for certain kinds of bursts or repetitive behaviors. If you are wondering why a mid-1990s satellite would use old proportional-counter technology, then you've been reading closely.

Although RXTE uses old technology, it possesses two advantages most other X-ray satellites do not have: a fast response time and a large data rate. The proposers of the project designed the internal electronics to send the collected X rays for onboard processing to the arithmetic chips at a high speed. This design provided the satellite with the capability to process data immediately after detection. The design also allowed limited programmability of the chips to respond to particular types of changes in the arrival rates of X rays.

The high data rate is perhaps the key difference between RXTE and other observatories. No X-ray source other than the Sun and Earth is too bright for RXTE.[36] Instrument builders impose brightness limits on observatories, such as Chandra and Newton, to prevent overloading the detectors. For Chandra, the data rate for science is about 25 kilobits per second (32 kilobits per second total). RXTE can handle data rates, for short times, up to one megabit per second. The high data rate plus the large effective area provide high signal-to-noise data or data with a high time resolution, both of which advance our understanding of the behavior of matter that occurs close to the surface of a neutron star or near the event horizon of a black hole. RXTE discoveries to date are truly impressive and support investigations of two of the fundamental questions of NASA's Office of Space Science "Structure and Evolution of the Universe" road map:[37] cycles of matter and energy in the evolving universe, and the extreme physics of high-gravity environments.

For example, around the neutron star Centaurus X-3, a team uncovered evidence for the existence of photon bubbles. Photon bubbles had been a theoretical construct, first predicted to exist about 30 years ago, with details worked out in the mid- and late 1980s.[38] Using supercomputers, theorists at Lawrence Livermore

National Laboratory studied the behavior of matter falling onto a neutron star from a companion. With a strong magnetic field, the matter will not fall just anywhere, but will be concentrated at the magnetic poles, forming an extremely energetic aurora perhaps similar to Earth's aurorae.[39] As the matter crashes into the neutron star, local regions of higher and lower density are created. Radiation from the hot matter floods into a pocket of slightly lower matter density. When the pocket collapses, the X rays escape all at once. RXTE recorded a flickering in the number of X rays from Cen X-3 that occurred about one thousand times a second; the formation and collapse of photon bubbles provide one possible explanation. We must obtain additional data to establish whether the theory of photon bubbles is correct. That an explanation such as photon bubbles is viable for neutron stars demonstrates that the space surrounding an accreting neutron star is bizarre.

Let's consider another odd source, XB1730-335, known as the Rapid Burster. Figure 5.6 shows a set of data obtained by the Italian X-ray satellite BeppoSAX. The data were obtained during an outburst in February 1998. The behavior of the Rapid Burster in 1998 matches the behavior it exhibited when first discovered in the 1970s, so it has not changed during that time. Look closely at the horizontal

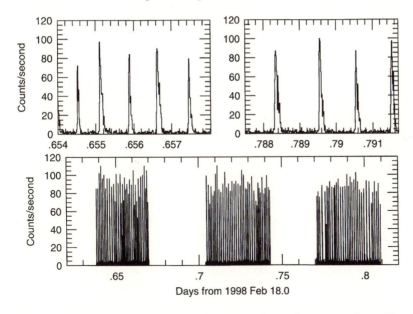

Fig. 5.6. An outburst of the X-ray binary XB1730-335, or the Rapid Burster, as observed by the Italian X-ray satellite BeppoSAX (named for the "father" of Italian X-ray astronomy, Giuseppe ["Beppo"] Occhialini). The lower half of the figure shows a portion of the complete light curve obtained about 20 days after the start of the outburst. The gaps in the light curve occur at times when the source is not in the field of view of the detector because Earth blocks the source's light. The upper two plots expand portions of the lower plot to show the regularity of the bursts. Each burst lasts about 15 seconds and recurs about every two minutes. (Plot generated from data obtained from the SAX public data archive at the High Energy Astrophysics Science Archive Research Center, NASA-Goddard Space Flight Center. Similar figure in Masetti et al. [2000].)

scale, the time axis. The divisions are fractions of a day; the plots show that the bursts occur essentially about every 15 seconds, and there are about 40 to 45 bursts per hour. During a burst, the observed count rate increases from about a few counts per second to about 80 to 100 counts per second. Over a longer time span, the Rapid Burster enters an outburst phase lasting several weeks followed by a quiescent interval lasting about six months.

The Rapid Burster is a low-mass X-ray binary living in or near the center of a globular cluster, a tightly packed group of stars resembling a miniature elliptical galaxy. The repetitive behavior suggests a spillover type of accretion in which a reservoir fills and overflows, with the spilled matter landing onto the surface of a neutron star and flashing as the matter rapidly heats. Until the mid-1990s, the Rapid Burster was a unique object; a second, recently discovered object, GRO J1744-28, shows similar behavior. In spite of the relative wealth of data on these sources, astronomers understand neither in detail. Particularly puzzling for X-ray-binary researchers is the paucity of other sources demonstrating the same behavior.[40]

Neither Chandra nor Newton is optimized for studying high-data-rate sources, so both might be expected to contribute little additional information to the study of X-ray binaries. Although the larger collecting area of RXTE and the higher rate at which it processes arriving X rays cannot be matched by either of the other observatories, RXTE does not have imaging detectors. The higher spatial resolution of Chandra and Newton allows both observatories to avoid contaminating sources.

Chandra, for example, has already contributed an exciting observation of a pulsar. At the center of the Crab Nebula, a remnant of a star that exploded in A.D. 1054, lies a pulsar. The Crab Nebula is the archetype for remnants of supernovae in which a pulsar exists. The pulsar, a neutron star spinning about 30 times per second, supplies energy to the remnant by slowly spinning down. Figure 5.7 shows the image of the pulsar. We see rings of matter that must be spinning with the neutron star. Streaks, or waves are visible in the disk. Flowing out from the center of the disk is a jet of gas. We do not really understand the details of this image precisely because it is a static image. One of the first observations carried out by the astrophysicist J. Hester of Arizona State University and his collaborators used Chandra (and Hubble) in a short series of simultaneous exposures to study the dynamics of the wisps and knots of emission observed in previous studies. Hester and his colleagues used the Hubble and ROSAT in the early 1990s to carry out a similar study. From that study, Hester learned that the wisps, moving outward at about half the speed of light, form and dissipate within about three weeks. The Chandra data provide higher spatial resolution, showing, for example, matter flowing from the jet as well as the formation and dissipation of the wisps. The pulsar is magnetized, so the team interprets the data as an exercise in the dynamics of electromagnetism at high temperature, high matter density, and high magnetic-field strength. Someday we will understand how the rotation of a pulsar supplies energy to the rest of the nebula.

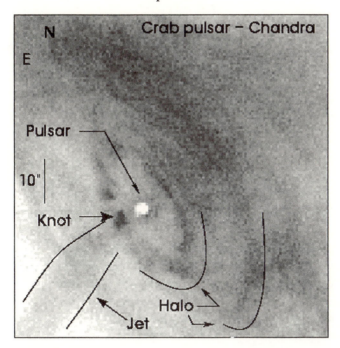

Fig. 5.7.   The Chandra observation of the pulsar at the center of the Crab Nebula. The white square is the overexposed image of the pulsar that was excised prior to printing the image. Several of the features in the nebula are indicated. A still image such as this does not do the observation justice; a short movie shows the magnetohydrodynamic flow present in this source. The swirls are the waves of plasma moving away from the pulsar. (Image generated from data obtained from the Chandra public data archive at the Smithsonian Astrophysical Observatory.)

The title of this book is *The Restless Universe*; the reader might perceive the restlessness from the discussion in this chapter. The ASM is not an imaging detector, but from its data, a cartoon image may be created. The ASM team at MIT has done that, producing a movie of about three years of ASM data with dots of various sizes to represent the different brightness of the various X-ray sources in our galaxy.[41] The restlessness of the X-ray sky becomes immediately visible as the X-ray binaries in our galaxy resemble a line of huddled people constantly shuffling to keep warm. The producers of the movie color-coded the dots to indicate the spectral behavior of the sources; spectroscopy is the subject of the next chapter.

6

"... a spectrum is worth
a thousand pictures."

—Unknown astrophysicist

Vibrant pictures delight the eye. We've all heard the phrase "A picture is worth a thousand words." To this, an astrophysicist has added: "But a spectrum is worth a thousand pictures."[1] Let's see why.

First, let's agree that pictures are seldom quantitative, because they have not been calibrated.[2] This does not mean that the images themselves do not convey measurable information. For example, if we know the distance to the imaged object, then we can measure the object's size. Sizes and relative brightnesses obtained from an image may be a prelude to a more detailed study.

The image of the plumes of star formation in the Eagle Nebula obtained by the Hubble Space Telescope is fabulous for its detail and justifiably one of the more famous images from Hubble. The images of the nebula, sculpted by winds from newly forming stars, resemble imaginative paintings in a New Age gallery. Few in my experience have looked at the plumes without saying "Wow." Professionals are no different: we become just as excited by the images. But we want more information. How fast is the gas expanding or contracting? What gases are present? Does the object have a magnetic field? Is it rotating or moving around an unseen body? What is the temperature of the gas? Is the gas optically opaque or transparent?

So we ask about the spectrum. What does it look like?

A colorful picture is useful because it tells us where to aim the spectrograph. For example, in the Eagle Nebula, astrophysicists definitely want to obtain a spectrum of the bright edges of the plumes: the gas is brighter there than at other locations. Why? What does the spectrum look like? The image of a bright edge of a plume tells the observer where to point the telescope to obtain a spectrum.

Spectroscopy is often glossed over in presentations to a general audience because it is perceived to be difficult to understand without a detailed knowledge of physics. This chapter discusses spectroscopy and why it is so important to astrophysicists. Let's start with a crude analogy.

Consider a politician. She wants to be a good politician, so she must strike a balance between following her interests and meeting the needs of her constituents. Let's assume this politician is an honest one, so that being a good politician means listening to her constituents closely and representing them as forthrightly as she can. If our politician wants to attain a higher office, or if she simply wants to keep her job, she must understand what her supporters want. She must assess the level of support she has from constituents because her ability to meet their desires depends directly on their support. Our politician hires a pollster to conduct a poll.[3] The pollster asks a single question of each survey participant: "Do you like politician X?" or "Is politician X doing a good job?" The answers provide our politician with a measure of her support.

Measuring the number of people who support politician X is equivalent to photometry. In other words, it is a simple measure: in the politician's case, of her support; in the astrophysicist's case, of the brightness of a star. That number can be followed over time, so our politician will know how her support waxes and wanes. A plot of brightness versus time is known as a light curve. What does it tell us? From the light curve we learn whether an astrophysical source, be it a star, a planet, or a galaxy, is constant in time or variable. Except under special circumstances, the amount of information contained in a light curve is not large, nor is it large for our politician if all she learns is the level of her support.

How else can she learn about who supports her and who does not? She orders her pollster to ask additional questions. For example, suppose the poll asks the same question as before ("Do you think politician X is doing a good job?") but the pollster also records the age of the person providing the answer. Now, our politician may find that people younger than 35 support her, but those over 55 do not. Our politician has learned something fundamental that can aid her in deciding where to focus her campaign.

Once our politician sees the value in asking additional questions, the questionnaire will expand faster than the Big Bang. Questions will be added to define the nature and interests of the respondent: what is your educational, religious, income, and ethnic background? What was the educational level of your parents and grandparents? How many children do you have or do you want? How often have you moved? What sports do you enjoy playing? Watching? What movies do you like? How much money did you spend last year in restaurants? The list is essentially endless.

Pollsters must ask questions such as these of a sufficiently large group of people to construct an array of answers for a statistically meaningful sample. Pollsters and politicians call such detailed descriptions of survey respondents "voter profiles" or perhaps a "profile of the constituency." An astrophysicist might say that the politician has obtained a spectrum of the voters.

The constituency profile contains the detailed picture of who supports our politician on which issues. The expanded poll provides the politician with the details of her constituency; she can predict how they will react to any particular message she distributes.

Spectra of astrophysical sources serve the astrophysicist much as a poll serves a politician. Spectra contain the detailed clues that often unlock whatever puzzle the source presents to us. Those clues might consist of gas motions, gas temperatures and densities, or perhaps the presence of a magnetic field. Ultimately, that is why we spend hours collecting the data and analyzing it. We need the spectroscopic data to obtain the clues to understand the physics of the object under study. The detailed clues turn "stamp collecting," in other words, collecting pictures of astronomical objects, into astrophysics. It is the astrophysical understanding that lies at the heart of our efforts.

Picket-Fence Spectroscopy

Visualize a picket fence with a light behind it.[4] The light can be sunlight or a streetlight; its nature is irrelevant. What do you see? You see bright regions alternating with dark regions (Fig. 6.1, top). Photons from the light travel to your eye except where the slats in the fence block the light. The entire length of fence represents the electromagnetic spectrum. The dark regions represent absorption lines. Absorption lines are locations in the electromagnetic spectrum where atoms of gas have removed, or, more precisely, absorbed, the light before it reaches your eye (or, for astrophysicists,[5] the detector).

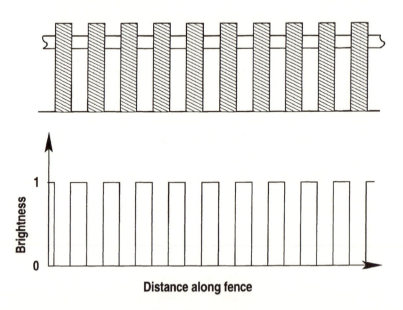

Fig. 6.1.   A picket fence (top) and its accompanying spectrum (bottom). The spectrum shows the intensity of the light at a given energy plotted on the vertical axis; length along the fence (equivalent to energy or wavelength or frequency) is plotted on the horizontal axis.

Suppose we construct a graph of this spectrum. Let the horizontal axis represent distance along the fence (in other words, frequency or wavelength in the electro-magnetic spectrum) and the vertical axis represent the intensity or brightness at a particular point on the fence (in other words, at a particular frequency). A graph of the light shining through the fence then resembles Figure 6.1 (bottom). The actual light might be from a star or a cloud of gas or some other astrophysical source.

When we ask a contractor to build a picket fence, we expect a fence with even spaces between one slat and the next. To make our analogy more realistic, how-ever, visualize a fence in which some of the slats are wide while others are narrow, and some of the gaps are wide while others are narrow (see Fig. 6.2). Let's explore some extreme situations. First, imagine the width of the slats shrinking every-where to zero. No slats (atoms of gas) interrupt the light reaching our eyes, so the brightness will be continuous from one end of the fence (one end of the spec-trum) to the other. This spectrum is called a continuous spectrum, or the con-tinuum.[6] How relevant is this? You, the reader, being of sound mind and not deceased, emit a continuous spectrum with a temperature of about 98.6 degrees Fahrenheit (37 degrees Celsius, or 310 degrees Kelvin). An object with that tem-perature emits radiation that peaks in a band called the mid-infrared. The peak in the mid-infrared does not mean that all the radiation is emitted only at those wavelengths. A continuum means just that: radiation is emitted across a broad range of wavelengths.

Under certain conditions, our continuous spectrum will be a Planck spectrum (Fig. 6.3 [see next page]).[7] Any solid object with a temperature above absolute zero emits a Planck spectrum. A single quantity identifies a Planck spectrum: the tem-

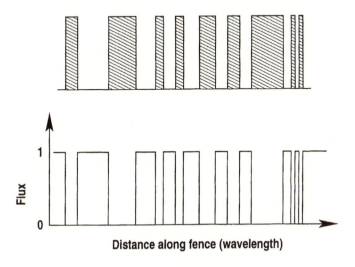

Fig. 6.2. A picket fence with uneven slat widths and separations (top) and the spectrum produced by the fence (bottom).

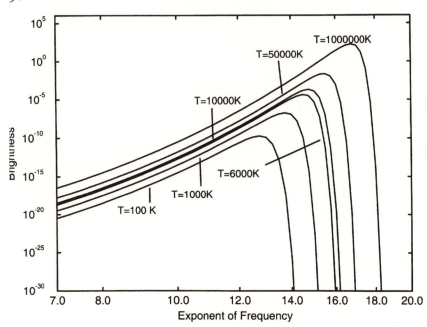

Fig. 6.3.   A series of Planck spectra. These plots use the same axes as the previous spectra, except the horizontal axis plots the wavelength of the light instead of the length along the fence. Each curve is labeled with a temperature. The unique aspect of the Planck spectrum: its shape is dictated solely by the temperature. Note that for higher temperatures, the peak of the curve shifts to shorter wavelengths. Note also that for a higher temperature, the entire curve lies at higher intensities at all wavelengths. In other words, as the temperature increases, the intensity at all wavelengths increases.

perature of the hot body. This is valuable because if we can show that an object, regardless of its location, is dense and hot, we know its spectrum just by measuring a temperature. In astrophysics, we invert this process: given a series of brightness measurements at different wavelengths, we match the measured brightness values to a Planck spectrum and derive a temperature. In practice, stars do not emit perfect Planck spectra. Instead, we assume they do and obtain an approximate temperature. For some stars, the approximation is rather good; for others, it is poor. Let's separate the stars based not only on their approximate temperatures, but also on how well the Planck spectrum describes the stellar spectrum.

There are basically three ways for a spectrum to be poorly matched by or deviate from a Planck curve: first, the actual spectrum falls below the Planck curve at one or more points; second, the spectrum lies above the Planck curve at one or more points; and, third, the actual spectrum shows no curve at all. How can we understand these cases?

Just above, we imagined the width of the fence slats shrinking to zero. If we now imagine the gaps between the slats shrinking toward zero (but not reaching it), what do we see? Each gap represents a piece of the spectrum that reaches our eyes. If these gaps shrink, then the slats (absorption lines) absorb most of the spectrum. Imagine that each absorption line represents energy absorbed, or removed, by atoms of a particular gas. Because we are conducting a thought experiment,[8] we can adopt this restricted viewpoint for the moment. We'll revise it shortly to match reality. In the context of our thoughts, what have we learned? If we know which gas produces a particular absorption line, then we can look to see if that absorption line is present. If it is, we know that the corresponding gas lies somewhere between the star and us.

Let's skip the second situation for just a moment and consider the third: the spectrum shows no Planck curve at all. Instead, there is just a series of bright lines. How do we understand this? For the moment, set aside our picket-fence model. Instead, stretch out a string of colored lights. Once again, distance measured along the string of bulbs corresponds to wavelength or frequency. Colored Christmas lights are evenly spaced on electric cords. Let's alter that so the spacing between bulbs is uneven. Furthermore, let's rearrange the colors so that the blue bulbs are at one end, the reds at the other. When illuminated, the string represents our line spectrum. Adopt the same model as we did for the absorption lines. Now, however, the atoms of gas emit radiation, creating the bright lines.

We bypassed the second situation for a short time; let's return to it. The second case occurred when the spectrum fell above the Planck curve at one or more points. In other words, extra emission existed at one or more wavelengths. We understand this case by placing the string of lights near the picket fence in such a manner that at one angle, the string lies in front of the picket fence, while at another angle, the string lies silhouetted against a dark background.

When we look at our fence at the specified angle, we see the continuum light and we see the places where the slats block the radiation, but we also see a few locations where extra light is present. The extra light represents additional emission. At a different angle, we see just the string of lights. The string of lights represents gas lying in the vicinity of our "star." The emission spectrum represents radiation at a few selected wavelengths. Again, the bright lines tell us which gases are present.

Under what physical conditions do we see absorption lines? The gas must be sufficiently dense to absorb enough light from the continuum for us to detect the absorption line. It must lie along the line of sight between the continuum source and us.[9] Finally, the gas must be cooler than the source of the continuum.

The density requirement is easy to understand. If the gas is too rarefied, the light will pass through it without encountering any atoms of gas, so no absorption will occur. For our eyes or instruments to record any absorption, the gas doing the absorbing must fall between the continuum source and us. That's sensible. But how do we know that the gas must be cooler?

The flow of energy from hot to cold is one of the fundamental behaviors of the universe. For the gas to absorb energy, the atoms must be cooler than the source of the energy. By this I do not mean that the gas must be physically separated. The outer layers of a star, for example, are still part of the star, but they are cooler than the inner layers. The outer layers absorb radiation emitted by the hot core. The absorbed energy still escapes the star, but it does so at some other wavelength.

The presence of absorption lines means the star no longer emits a Planck spectrum. Too little energy escapes in the band where the absorption lines exist; too much energy escapes in some other spectral region. If the absorption lines block only a small portion of the escaping radiation, then the approximation of the spectrum as a Planck spectrum will be good. The solar spectrum is rather close to a Planck spectrum at a temperature of 6,800 degrees. Absorption lines intrude, particularly in the blue portion of the spectrum. Stars cooler than the Sun show considerably more absorption lines, while stars hotter than the Sun show increasingly fewer lines.

One word of caution: real spectra seldom fall into the nice, neat categories of Planck, absorption, emission, or picket-fence spectra. Real spectra, emitted from real stars or gas clouds or active galaxies or supernova remnants, often combine all the nice, neat situations laid out above.[10] Suppose, for example, our picket fence has a few slats that glow, some slats that are wide, some that are narrow, a few gaps that are wide, and a few that are narrow. Then the fence and the graph might look like Figure 6.4. This begins to get messy, doesn't it? Welcome to spectroscopy!

The picket fence and the light-bulb string are models for absorption and emission spectra. How do these models fail to describe reality? In using these models, we have made several assumptions, or approximations to reality. The approximations allowed us to move forward with the discussion, but we should return to the approximations to see if they should be relaxed or eliminated.

First, in the picket-fence model, we assumed that a slat blocks the light completely, whereas an atom seldom absorbs radiation completely. To alter our picture, replace the slats with increasingly densely smoked glass, for example. The shape of the absorption line might then resemble Figure 6.5. Even at the center of the absorption line, a small quantity of radiation may escape the star. The edges of the line are not sharp but occupy a greater number of wavelengths near the level

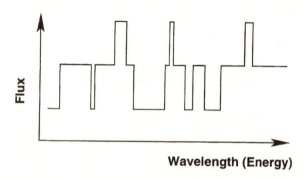

Fig. 6.4.   A "realistic" picket-fence model with absorption and emission lines on a continuum.

Fig. 6.5. A picket fence with a single slat that becomes increasingly opaque, then increasingly transparent.

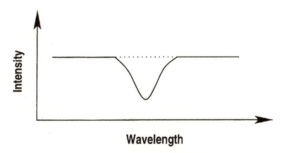

of the continuum. We'll see shortly the importance of that behavior. The use of slats suggests a thin layer in a star that absorbs the light; in reality, absorption occurs along a path toward the observer. Second, we also approximated emission lines as bright slats emitting radiation at a few specific wavelengths. In reality, an emission line resembles Figure 6.5 inverted. Finally, our picket-fence model assumed that the gas was purely of one element. The next chapter will relax this assumption and examine its implications.

Figure 6.6 shows an optical absorption spectrum of a white dwarf. White dwarfs were once stars like the Sun. A star exists because of the opposing forces of gravity and pressure. Gravity pulls the star's layers inward; gas and radiation pressure push the star's layers outward. The pressure exists only as long as a star has fuel to

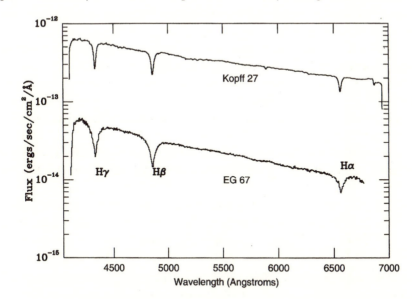

Fig. 6.6. *Upper:* The optical spectrum of a main sequence A star. *Lower:* The optical spectrum of a white dwarf in the visible portion of the spectrum (a measurement of 4,500 angstroms corresponds to violet/blue light; one of 6,500 angstroms corresponds to red light). Note that the continuum for both stars follows a smooth trend, but is interrupted by absorption lines. These lines identify the absorbing gas as hydrogen. If the complete continuum were visible, it would resemble a Planck spectrum. The two stars differ by a factor of ~100 in size and therefore in luminosity by 10,000 or more.

burn in the nuclear reactions that occur in and around the star's core. Once the fuel runs low, the outward pressure starts to drop, and the outer layers fall inward. This is a runaway catastrophe for massive stars. For low-mass stars, however, the inward fall does not lead to catastrophe. Instead, the fall causes the gas temperature to increase, which increases the gas pressure. Gradually, however, gravity forces the atoms in the core into a state known as "degenerate matter."[11] At this point, the star is a white dwarf. It is essentially dead because nuclear reactions no longer occur in the core. It will continue to shine for many billions of years as it gradually cools off. The collapse of the core of a massive star also causes the core temperature to increase, just as in the low-mass case. However, the resistance to collapse offered by degenerate matter in the low-mass case is woefully insufficient to halt the core collapse for high-mass stars.

Back to the white-dwarf spectrum. Look at the shapes of the absorption lines. Each looks identical to the others except for depth and position. This is a textbook absorption spectrum.[12] The spectrum below the white dwarf is a spectrum of an A star. A stars have surface temperatures of about 10,000 K. The spectra look the same, yet these stars could not be more different. The A star is still using its hydrogen fuel, happily converting it to helium in the nuclear reactions in its core. How do we know these two stars are so different? We know only if we can estimate the distance to the star and thereby calculate the luminosity. The luminosity of the white dwarf is about a hundredth that of the A star; its radius is about a hundredth that of the Sun, or about the size of Earth.

We extract other quantities from these spectra. The detailed shape of the line, when compared to theoretical line shapes, indicates the pressure of the atmosphere.[13] If the star is moving, the absorption lines will be displaced either to the blue (star approaching us) or to the red (star receding). This impressed motion results from the Doppler shift, the same process by which sounds change pitch as the source approaches or recedes from us.

Light behaves as a wave.[14] When a moving atom emits light, the frequency of the light is shifted to either lower or higher values, depending on the direction in which the atom is moving relative to the observer. If the atom is moving away from the observer, we see the frequency shifted to lower values (equivalently, to lower energies or longer wavelengths; shifts to lower frequencies are known as redshifts); the reverse occurs for atoms moving toward the observer (blueshifts). This shift in frequency is known as a Doppler shift, first identified by Christian Doppler in 1842. Sound waves also undergo Doppler shifts. The classic example is a train whistle, but sirens and lightly muffled engines, such as that of a motorcycle, also work. Listen closely for the change in pitch the next time a distinct sound approaches you. The size of the pitch change increases for higher speeds. The sound changes from a high pitch to a lower one as the vehicle passes. In astrophysical spectra, a Doppler shift changes the frequency (or the wavelength) at which we find the emission (or absorption) line.

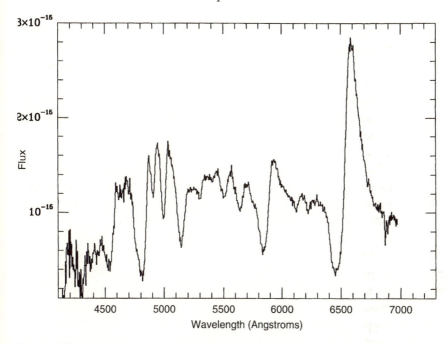

Fig. 6.7. The optical spectrum of a supernova (SN1988H). Note how the lines blend together in places, so much so that it is difficult to define a "Planck continuum."

Figure 6.7 shows an optical spectrum of a supernova,[15] a star that has exploded. This spectrum shows emission lines, absorption lines, and a continuum. Remember the picket-fence model. Try to pick out the continuum, absorption lines, and emission lines. Interpreting such a spectrum challenges the ingenuity of astrophysicists. But it's fun.

The supernova spectrum also illustrates another difference between the picket-fence model and a real spectrum: in a real spectrum, neighboring lines can blend together. In the supernova spectrum, a blend of an absorption line and an emission line is visible at about 6,600 angstroms: a line of hydrogen. The line shape indicates that material lies between the center of the supernova (the absorption) and us, but material also has been excited so that it glows on its own (the emission). Furthermore, because both lines are hydrogen, the absorbing matter and the emitting matter must be physically separated in some manner because the centers of the two lines are offset from each other.[16] The blending of many absorption lines acts essentially like a blanket, trapping energy within the star; the trapped energy eventually emerges, usually at a longer wavelength.

Why the focus on optical spectra? To illustrate two points: first, the signal-to-noise ratio is higher with optical spectra, so the picket-fence model is more easily compared to real spectra; second, the entire discussion applies to any spectrum,

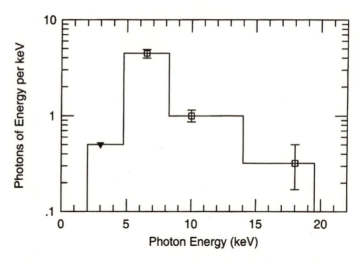

Fig. 6.8.   One of the earliest X-ray spectra obtained by a rocket in flight in October 1964. The spectral resolution here is about 50 percent. (Plot reproduced by permission of Dr. Philip Fisher; the original plot appeared in *Astrophysical Journal* 143 [1966]: 203.)

regardless of wavelength. Once you've learned the basics of spectroscopy, that knowledge can be applied at any wavelength.

So, let's look at some X-ray spectra. Figure 6.8 shows the spectrum of Sco X-1 from a rocket flight in 1964.[17] The spectral resolution is minimal; we essentially have three or four samples of the X-ray continuum. This spectrum was valuable in 1966 because it was one of the first X-ray spectra ever obtained. It illustrates, for example, that the continuum is declining toward higher energies. From a more recent satellite, Figure 6.9 (top) shows a spectrum with a temperature of 3 keV in the region of the iron line at 6.5 keV from the proportional counters on Rossi X-ray Timing Explorer. Note the difficulty in identifying whether a line even exists. The counters on RXTE have little spectral resolution; however, the bandpass covered by the counters is very large: from 2 keV to about 20 keV, a range of 10 in energy; this demonstrates one of the advantages of using RXTE. In the optical band, our eyes are sensitive from about 3,200 angstroms to about 7,500 angstroms. If our eyes covered a factor of 10 in wavelength, we'd be sensitive to light out to about 30,000 angstroms, which lies well into the infrared band.

Figure 6.9 (middle) shows the same spectrum based on the response of the CCD spectrometer on Chandra (ACIS). Figure 6.9 (bottom) is a spectrum with still better resolution; this uses the response of the high-energy-diffraction grating on Chandra. We see a vivid example of the value of increased spectral resolution. The three spectra are simulated data intended to match the emission from an interacting binary star. The CCD and proportional-counter spectra show blended emission lines even for a relatively isolated energy range such as the 6.5-keV region. Near 1 keV, however, emission lines from iron, neon, and silicon, among

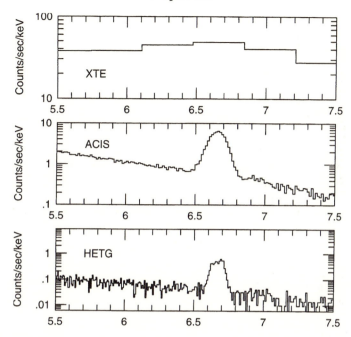

Fig. 6.9. Comparison of the spectral resolutions of the prime instruments on RXTE (the Proportional Counter Array) and Chandra ACIS and ACIS plus the High Energy Transmission Grating. The spectra compare the resolution in the 6.5-keV band, where emission lines from iron appear. Each spectrum shows the exact same model. From RXTE data, you would have difficulty stating unequivocally that an emission line exists; with Chandra, the emission line is readily apparent.

others, prevail. Poor spectral resolution easily blends these lines, so the physical interpretation becomes at best difficult and at worst error-prone. Every scientist dreams of obtaining spectra with superior spectral resolution. Most of the spectroscopy that has been available to X-ray astronomers resembles Figure 6.9 (top) or Figure 6.9 (middle). Chandra and Newton are the first for which grating spectra will be routinely available for a wide range of sources. A grating spectrometer existed on Einstein, but the effective area was sufficiently small that astronomers could study only one or two very bright objects.

The Chandra Detector

Just how is the energy of an X ray determined? One of Chandra's detectors uses CCDs (as do the detectors on Newton). The CCD pixel measures the energy of an incoming X ray when the X ray interacts with the layers of silicon that make up the chip. The interaction liberates electrons in direct proportion to the energy of the incoming X ray. Once the electronics transfer the charge to the readout, the total charge is read. The statistical accuracy of the readout determines the energy resolution; currently, for CCDs, the energy resolution is a few percent for 3-keV X-rays.[18]

The spectral resolution of the detectors has increased over the past decade, leading to the first studies of line shapes in the X-ray band. Previously, lines were simply detected, but they were equivalent to lines with zero width. Physically this is not possible, because every line has a natural breadth. The poor spectral resolution, however, artificially broadens the line—artificially—because the detector is simply not capable of better resolution. Knowing that a line is present is useful because we know what gases are present, but we cannot investigate line shape.

Recent improvements in energy resolution have led to at least one significant discovery using the CCD detectors. Some lines remain unresolved, whereas the detectors determined other lines to be broad. Not only are the emission lines of iron in the active galaxies MCG-6-30-15 and NGC 3516 broad, but the line shapes may indicate the presence of black holes. Figure 6.10 shows the portion of the spectrum from NGC 3516 around 6 keV near the iron line.[19] The Japanese-American satellite ASCA obtained the spectrum from an exposure that lasted, once the

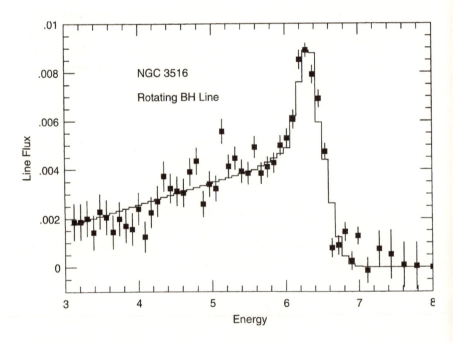

Fig. 6.10.   The iron-emission line from the active galaxy NGC 3516 as observed by ASCA's CCD detectors. The thin solid line drawn through the data is a model that assumes that all the emission arises from an accretion disk surrounding a galactic-sized black hole. The match of model and data is impressive. The emission-line feature was first reported by P. Nandra. (Plot reproduced by permission of Dr. Paul Nandra, NASA-Goddard Space Flight Center/Universities Space Research Association; the original plot appeared in Nandra et al., *Astrophysical Journal* 523 [1999]: L17.)

blocking of the source by Earth was removed from the data, for about two continuous days. The data are noisy because the source is weak;[20] the scatter indicates the photon noise.[21] Look at the shape of the line: there is a steep drop on the high-energy side and a gradual slope to the low-energy side. The line drawn through the data points is a prediction of the line shape, assuming the line is emitted by gas near the event horizon of a black hole. The line matches the data quite well.

Around a black hole, the emitting gas suffers the combined effects of the gravitational redshift of the black hole and a Doppler shift because the gas moves.[22] The gravitational redshift occurs as light leaves a gravity well. Light leaving the gravity well of any object with mass will show the shift. Its ease of detection depends on the mass of the object; being detected is easier for higher-mass objects. Physicists detected the shift in Earth's gravity field at a Harvard University laboratory in the 1960s; the astronomer D. Popper reliably detected the shift in a white dwarf in 1954. The Doppler shift occurs from gas moving at relativistic speeds (>10 percent of the speed of light).

Both effects distort a symmetric line shape. In fact, the line shape makes more sense if the model comes from a rotating black hole. The data are consistent with the line from a nonrotating black hole, but some of the model's parameters do not make sense within our understanding of astrophysical behavior. Now, the figure and the model do not prove that black holes exist, because it may be possible for some other process to produce a similar line shape. However, the evidence is strongly suggestive. How do we clinch the discovery? Two approaches may provide the smoking gun: still higher spectral resolution and images of the central sources. The next section addresses one path to improvements in spectral resolution, while another approach will be presented in the next chapter; imaging the central object will be covered in chapter 10.

The Detector Plus Gratings

By placing a diffraction grating into the X-ray beam, we obtain higher-resolution spectra. Almost everyone either has split sunlight by using a prism or has seen a rainbow, which is nature's way of dispersing sunlight into a spectrum. Two different concepts exist for breaking light into its component colors. Dispersion of light occurs because the speed at which light travels in the dispersing medium is different at different frequencies. The words "in the dispersing medium" are key: the familiar value of 299,792 kilometers per second (186,200 miles per second) is the speed of light in a vacuum. As a sunbeam passes through a rain droplet, the blue light moves more slowly than the red light. As a result, the colors take slightly different paths through the raindrop,[23] and upon exiting, physically spread out. A prism works in a similar manner.

Prisms and rainbows are familiar, so they are the natural entry point for the discussion. Diffraction gratings, however, work not by dispersion, but by diffraction. Waves diffract, or spread out, after passing through a narrow opening; ocean waves diffract if they pass through a relatively narrow channel.[24] Cut two openings and the result will be constructive and destructive interference. At some locations, the waves will reinforce each other (constructive); at others, the waves will cancel (destructive). The complete picture will show zones of waves (regions of constructive interference) interspersed with "dead" regions in which no waves exist (areas of destructive interference; each zone is called an "order").

Gratings are finely ruled pieces of metal. Gratings used for optical spectroscopy have up to about a hundred thousand grooves per inch of grating. The spectral resolution depends on the number of narrow openings; the more present, the higher the resolution. What does a grating look like? Are there "real world" examples of diffraction gratings? Absolutely. Take out your favorite compact disk and hold it so that light from a lamp reflects off the disk. If you tip the disk at the correct angle (about 45 degrees from reflecting the lamp's light directly into your eye), you will see a rainbow of color. That rainbow is the diffracted spectrum of the source of light illuminating the disk. The compact disk acts like a reflection diffraction grating.

Why does a compact disk behave like a diffraction grating? Compact disks have narrow tracks of raised bumps that digitally encode the recorded music or video.[25] Those tracks, because they are evenly spaced (about 1,600 nanometers apart), diffract light. If we think about a narrow pie slice taken from the disk, there are about 20,000 lines across the pie slice, which is about an inch and a quarter from tip to curve (about 32 millimeters). That means the average compact disk has about 16,000 grooves per inch. Diffraction gratings for high-resolution astrophysics often contain about 5 to 10 times as many lines per inch. As a result, the color you see reflected from the compact disk is at least 5 to 10 times less sharp than that from a grating built for astrophysics.

How do grating lines increase the spectral resolution of a detector? A diffraction grating spreads the light by a known amount. It essentially converts the goal of measuring the energy of an X ray into the goal of measuring the distance from the location of the undiffracted light. Measuring a separation between two locations is considerably easier and more accurate than measuring the energy of the X ray directly. Spreading the light yields greater spectral resolution. Where the best CCDs measure an X ray's energy to plus or minus 50 eV and appear inherently limited to that value, diffraction gratings provide a resolution that depends on the number of lines cut into the grating. For the gratings on Chandra, the resolution varies from 0.4 to 77 eV.

There is a cost. By spreading the light, the signal-to-noise ratio at any given energy is lower. The effective area of a telescope with a grating inserted drops because the grating does not transmit all the light; the grating also sends light into all different orders. A grating can be designed to send as much light as possible to a particular order; as we have seen, however, no additional object introduced into the instrument is ever 100 percent efficient, so light is lost. The gratings on Chandra and Newton provide the largest effective areas for gratings ever flown. As a result, many more sources are observable than have been so previously. The future, however, lies with a different technique.

A brief aside: gratings do help Chandra observations of bright objects avoid a particularly nasty instrument problem: photon pileup. As described earlier, a CCD generates a charge directly proportional to the energy of the X ray. That is fine, but we must read the CCD, using the "bucket brigade," to know the value of that charge. If another X ray falls into the same pixel before we have read the initial charge in that pixel, then the second X ray will also generate a charge proportional to its energy. The charge from the first X ray, already in the pixel, and the charge generated by the second X ray are combined. If we now read the CCD, we get the sum and we no longer know whether one or more X rays generated the charge. Avoiding this problem can be easy: observe only faint sources, because the probability of two X rays landing in the same pixel within the short time it takes to read the CCD is very small for a faint source. For bright sources, that probability can be quite high, and for very bright sources, there is no longer probability but, instead, certainty. This is actually a drawback to the superb image quality of Chandra. Pileup increases for sharper point-spread functions. The goal—obtain better spatial resolution so you can really see the origination of X rays—is the very thing that frustrates you by increasing X-ray pileup. Pileup also distorts the spectrum. A grating disperses the light across the entire CCD, so each pixel of the CCD responds only to an extremely narrow portion of the X-ray spectrum. The probability of pileup is considerably reduced.

X-ray Spectroscopy of Clusters

Earlier we learned that clusters of galaxies are hot. How do we know? We infer the temperature from the continuum shape of the spectrum. We've learned that clusters have temperatures of a few keV; some clusters are hotter, some are cooler. Clusters are physically large in the sky, sufficiently so that we improve our understanding with detectors of even modest spatial resolution (such as ROSAT) by analyzing the data from a cluster as if it were an archery target. Divide the cluster into rings or annuli, accumulate a spectrum for each annulus, and measure the temperature outward from the center. With a series of temperatures at different radii from the center, we measure a temperature gradient. If the gas in a cluster is hot, we naively expect the temperature to decrease as we move toward the cluster edge: a positive temperature gradient.

One surprise, first discovered in the 1970s by Dr. Peter Serlemitsos of the NASA-Goddard Space Flight Center, is the existence of a reversed or negative temperature gradient in the centers of some clusters. Within a specific radius that differs from cluster to cluster, the temperature actually decreases as we get closer to the center of the cluster. This behavior has been dubbed "cooling flow." The "cooling" part is obvious; the "flow" part comes from the behavior of gas in a gravity field without a visible means of support: it falls inward. Using ROSAT data, astronomers have argued that some clusters possess cooling flows with hundreds to thousands of solar masses falling into the cluster center each year.

This is a problem. No evidence for the flow in other wavelength bands exists. If hundreds of solar masses of matter are assembling at the cluster center each year, then, in a short time, observational signatures of that material should be detectable. New stars should be forming, or existing stars should illuminate the gas. To date, astronomers have not found any such signatures. Observations point out the likely culprit: the higher-resolution Chandra and Newton data show the presence of cooling flows, but huge volumes of cooling matter do not exist. The ROSAT data led to overestimates for the cooling masses of gas. Chandra data imply that cooling flows exist, but that they contain at most a few tens of solar masses of matter.[26] Such a small quantity easily flows to the cluster center and remains undetected because its signature at other wavelengths is still too weak.

Future Spectroscopy

Let's return to the comment that gratings are probably not the future of X-ray spectroscopy. Gratings spread the light out, which is undesirable; they are used because they are relatively inexpensive. CCDs measure an X ray's energy by direct interaction of the X ray with silicon, but they lack high spectral resolution. A better instrument would provide at least the resolution of a grating and measure the energy directly, as a CCD does.

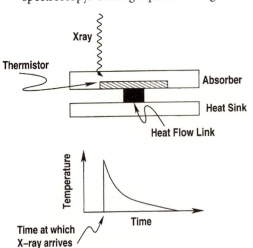

The newest approach, and the one likely to dominate X-ray spectroscopy over the next decade or two, is best described as an X-ray thermometer. The instrument is known as an X-ray calorimeter (Fig. 6.11). It works by absorbing the X ray, the energy of which heats an absorbing material.[27] The temperature of the absorber rises by an

Fig. 6.11.   A schematic diagram of a calorimeter (top) and its reaction after absorbing an X ray (bottom). A calorimeter was on board the ill-fated Astro-E observatory and was the key instrument for that observatory.

amount that is directly proportional to the energy of the X ray. Current passing through a thermistor, a semiconductor device that rapidly and predictably responds to temperature changes, translates the increase in temperature into a change in the voltage that the electronics detect. The absorbing material is critical: it must absorb the energy of the X ray as quickly as possible, yet have a low heat capacity; the temperature in a material with a high heat capacity will change immeasurably after absorbing an X ray.[28] The energy resolution depends on the noise in the absorbing material; cooling the detector to nearly absolute zero reduces the noise. In operation, spectral resolutions ten times better than CCDs have been demonstrated easily approaching or exceeding grating resolutions; resolutions of a few tenths of a percent are possible.

Calorimeters are just moving out of the lab. The first X-ray satellite to carry a calorimeter would have been the Japanese-American satellite Astro-E. The calorimeter on Astro-E had been measured to deliver a spectral resolution of about 10 eV at 6,000 eV (which is an energy resolution of about 0.2 percent). Figure 6.12 repeats the spectra shown in Figure 6.9, but with the addition of a simulated spectrum that the calorimeter on Astro-E would have delivered. The best CCDs deliver spectral resolutions of about 50 to 100 eV, or about 1 percent. The simulation shows that the resolution of the calorimeter meets or exceeds that of the gratings. Because of the nature of the measurement process, the instrument efficiency is relatively constant across the entire bandpass, in contrast to the efficiencies of CCDs or CCDs plus gratings, which typically drop with increasing energy. That constancy provides a distinct advantage to calorimeters.

At the 0.1 to 0.2 percent level of spectral resolution, astronomers measure gas motions if the gas moves at least 1,000 kilometers per second

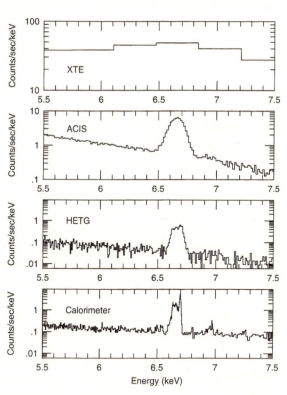

Fig. 6.12.   A plot comparing the spectral resolutions of different instruments, including the calorimeter that was to have flown on Astro-E. The other three spectra are identical to those in Fig. 6.9. The calorimeter data are simulated to match the temperature and absorption of the other three spectra.

(about 670 miles per second). Measuring gas velocities adds dynamic motion to our toolbox. Our interpretations of spectra to date essentially assume a static situation because, until recently, we have had little or no evidence for motion. We know that gas seldom sits statically because, for example, either radiation pushes it or gravity pulls it. To date, little evidence exists in observed spectra. As we resolve spectral lines, we first detect their breadth. We interpret line breadth as the motion of the gas.[29] With still higher resolution, we measure changes of position of the line, which is a direct measure of gas motion. Following the change in position of a line over time gives us orbital or rotational dynamics in X-ray binary stars, for example, which is an extremely important tool to help us understand the detailed processes of X-ray emission.

Spectra obtained with Astro-E would have given us our first direct look at the dynamics of gas in binary-star systems containing accretion disks as well as the accretion disks of active galaxies and newly formed stars. The loss of Astro-E hurts for that reason.

The identification of patterns and regularities remains a goal of any observational science. Eyes exist on nearly every creature that swims, flies, or walks. That is a pattern because it puts light inside the dense bones that form the heads of every mammal. Nature did not use another available solution, namely, the radar or sonar system of bats, perhaps because, although it works for one mammal, if every mammal used it, chaos would reign.

An astrophysical equivalent of the eye is the accretion disk. Accretion disks create dynamic behavior when they are found in binary-star systems. The gas in an accretion disk circulates about the central object at speeds dictated by the mass of the central object. Velocities of 1,000 kilometers per second and more are easily achieved. Accretion disks surround newly formed stars, where they are believed to lead to the formation of planets. Disks also exist at the centers of active galaxies. A galaxy is active when its nucleus emits considerably more energy than the nucleus of a normal galaxy.

For active galaxies, however, the whole is considerably greater than the sum of the usual parts. To explain an active nucleus, astronomers invoke the presence of a massive black hole at the center of the galaxy. Such a black hole is sufficiently large that whole stars can be swallowed by it. The black hole may possess a mass that is about one million to a hundred million times as large as that of the Sun. Yet the manner in which these black holes behave may be similar to that of X-ray binaries. Jets of matter emanate from many active galaxies. Astronomers have also discovered jets of matter, operating on a smaller scale (say, the difference between a fire hose and a squirt gun?), from a few X-ray binaries. Astronomers call these objects "micro-quasars." Debates rage over the appropriateness of the label.

Several types of active galaxy exist. In the mid-1980s, R. Antonucci and J. Miller advanced the idea that the system orientation dictated the type of active galaxy observed.[30] Astronomers expect the accretion disk surrounding the core of a galaxy to take on the shape of a torus, or a thick disk. If we view the galaxy from an intermediate angle, neither along the disk-rotation axis nor perpendicular to it,[31] then we see broad optical emission lines; if our view is close to the equator, so that we look through the torus, then we see narrow optical emission lines. In the X-ray band, however, the two views differ more dramatically. The inclined viewpoint sees directly into the inner disk; the equator view shows the inclined view, but with heavy absorption. In a galaxy viewed near its equator, we see few X rays. The AGN viewpoint model received support with the detection by RXTE of the heavily absorbed active galaxy NGC 4945:[32] essentially no emission exists below about 10 keV other than X rays reflecting off gas surrounding the central object. Optically, NGC 4945 is an active galaxy; based on an X-ray spectrum in any band below 10 keV, the galaxy is not. RXTE's broad spectral coverage (2 to 200 keV) reveals the heavy absorption; the spectrum above 10 keV appears identical to that of other active galaxies. This result motivates one design for future observatories: increase the spectral resolution to see the lines more cleanly, but be certain to include a very broad range of energies. Chapter 10 will present an observatory on the drawing board that will address both requirements.

Recently, two teams of astronomers presented additional evidence in two separate publications that showed a correlation between the mass of the central object and the range in velocities of gas and stars surrounding the central object.[33] Both groups' results show that as the velocity range increases, the mass of the central object increases. If the addition of objects such as NGC 4258 confirms the correlation, then astronomers will use the correlation itself to estimate the masses of the central objects of active galaxies. The galaxies listed by the two groups all contain multimillion-solar-mass central objects, from the "lightweight" one in our galaxy, the Milky Way, with a center estimated as two to three million solar masses ($2-3\times10^6$), to the very active galaxy M87, the central galaxy of the Virgo cluster of galaxies, with a center estimated to be about three billion solar masses ($\sim3\times10^9$).

The reality of black holes at the centers of AGN received a substantial boost by the discovery of the NGC 4258 emission because the emitting sources in the disk provided a clear measurement free of extra assumptions. The emission lines from the accretion disk surrounding the active galaxy's black hole behave similarly to the emission lines in the accretion disk surrounding a white dwarf or neutron star. This means we can use the emission lines to probe the gravitational field of the black hole, and by doing so obtain its mass. The active galaxy NGC 4258 presents the best case; in the mid-1990s, emission in the microwave band from water vapor was discovered near the center of the nucleus. Subsequent study showed that the microwave emission revealed the bulk motion of the gas in an accretion disk surrounding the central object. The measured Doppler shifts correspond to velocities of about 1,000 kilometers per second (600 miles per second) and establish the existence of an accretion disk. The motion in the disk implies a central object with about 40 million solar masses.[34]

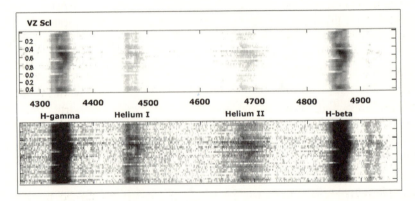

Fig. 6.13.　An optical image of the spectrum of an accretion disk of a cataclysmic variable. Wavelength in angstroms is across the page, time moves down the page, and the emission-line intensity is coded as shades of gray. The entire spectrum is shown twice at different contrasts to emphasize different features. The S-shaped pattern is the Doppler motion as the two stars revolve about each other. This illustrates the goal of increasing spectral resolution in the X-ray band.

How does the gas in an accretion disk behave? When looking at an accretion disk, the observer sees part of the disk moving toward, and part of the disk moving away from, the telescope, a natural result of gas undergoing rotation or circulation about a central axis. In the spectrum, the emission lines show Doppler shifts toward and away from the observer. The seemingly dual to-and-fro motion occurs because the observer obtains a snapshot of the disk. If the observer follows the motion of the gas long enough that the stars in the binary undergo one complete revolution about each other, then the observer will see a complete cycle of the motion of the gas.

Figure 6.13 shows the (optical) spectroscopic results for the cataclysmic binary star VZ Sculptoris. The apparent "double line" is the spectrum of a single emission line, but doubled because of the Doppler shift from the two edges of the accretion disk. The amount of the separation gives us a measure of the speed of rotation of the accretion disk. In the case presented here, the speed is about 350 kilometers per second (about 200 miles per second). The "S" shape occurs because the two stars are orbiting each other, so that sometimes the entire accretion disk itself is coming toward the observer and at other times it is moving away.

The optical behavior illustrated in Figure 6.13 remains a goal for X-ray astronomy. We'd very much like to have the X-ray observations to build a figure such as this.

Summary

In 1960, we knew next to nothing about X rays from astronomical sources other than the Sun. The rocket flight of 1962 enlarged our horizons. With Chandra and

Newton, following the successes of ROSAT, the space shuttle mission Astro-1 (which carried the X-ray telescope BBXRT), ASCA, and RXTE, we are receiving a solid look at the X-ray sky. The spectroscopy gleaned from solid-state detectors whetted our appetite for Chandra and Newton. We now know that some active galaxies show emission lines that resemble those expected from black holes. That knowledge would have been impossible without spectroscopy. Fancy pictures are nice, but they are not necessarily astrophysics. Again, pictures may be worth a thousand words, but a spectrum is worth a thousand pictures.

So if you want to show your astrophysical sophistication, the next time you see beautiful astronomical pictures, make a demand of the astrophysicist or the science reporter; paraphrase that famous line from the movie *Jerry Maguire*: "Show me the spectrum." You likely will get a startled reaction. Be warned, however. If you ask for the spectrum, you will have to sit through the explanation that follows. It may be long, but this is where the astrophysics really exists.

7

"We're all nothing but unified arrangements of atoms . . ."

—Alan Dean Foster

Science writers often describe atoms as little solar systems in which electrons take the place of planets in their orbits about the nucleus without clarifying the analogy. In the previous chapter, we learned about spectroscopy, in particular, emission and absorption lines. In this chapter, we tackle what the presence of emission and absorption lines tells us about the atoms that exist in the object under scrutiny and try a different approach.

Walk up a flight of stairs. Besides being good for your heart and leg muscles, stairs serve as a model of an atom. As you lift yourself from a lower step to a higher one, you expend energy. Physicists track energy just as accountants track money: the accounts must balance. For money, an account that does not balance may imply embezzlement; for energy, an account that does not balance is impossible.[1] As you step up to that next step, you lift yourself out of the gravitational field of Earth. This takes energy that your body supplies through the food you consume.

As you lift yourself up from one step to the next, you work against the force of gravity. Using the physicist's accounting, the energy you expend in going from one step to the next appears on the potential-energy side of the ledger. It is energy that can be tapped or used. If you could make one step vanish from under your feet, you'd fall vertically downward to the next step. (Ignore the small horizontal difference between one step and the next.) The energy stored on the potential side of the ledger would then be released as energy of motion, in this case, downward. You could use that energy to crush a soda can, for example. As you step down, you are executing what spectroscopists call a "downward transition." Turn around. As you step up, you execute an "upward transition."[2]

You can carry out a similar demonstration with a ball. Stand on the bottom step of a set of stairs and toss a ball upward. Try this several times, each time attempting to toss the ball upward with the same force. If you do so, the ball should reach the same stair. You can look at this from your perspective: you have supplied a certain force to propel the ball upward; or from the ball's perspective: it absorbed a certain amount of energy. As it bounces downward, from the ball's perspective, it is emitting energy.

Think of the ball as an electron; you represent the nucleus and the stairs the energy levels available to the electron. Following the toss, the ball will reach a certain height, then bounce back down. Using the terminology of atomic physics, you excited the electron away from the nucleus. Alternatively, the electron absorbed energy, namely, the energy you put into it during the toss. The electron reached a particular energy level dictated by the amount of energy absorbed. Changes from one level to another are called "transitions;" transitions occur between energy levels, either by absorbing energy for an upward transition or by emitting energy for a downward transition.

Notice that the steps are separated at fixed distances. To an atomic physicist, the stairs are quantized; they exist in discrete levels. The tossed ball can sit on any step, but not between steps. For an electron, the steps are called energy levels.

Our analogy is a crude one and is designed largely to apply atomic-physics terminology to an everyday experience. Examined in detail, it falls apart as a picture of a real atom because the classical or Newtonian world and the quantum world differ.[3] Most analogies break down if they are pushed too far; models used by scientists suffer the same problem. For example, Newton's laws of motion define a model of nature. For about three hundred years, the Newtonian model of nature proved robust. Albert Einstein showed that under certain conditions, Newton's model broke down and a new approach became necessary. Whenever a scientist discusses a model or analogy, you should always listen for its limitations.

Where does our stair analogy break down for a real atom? Stairs are quantized into constant units;[4] real atoms have quantized energy levels, but the difference in energy between levels is not constant. The ball bounces down from one step to the next; electrons may move down one or more energy levels in a single jump. The ball does not have a ground level below which it cannot sink, because if you dig a hole, the ball can fall into the hole. Electrons have a stopping point, the lowest energy level, otherwise known as the ground state; no energy levels exist below the ground state. Atoms possess a ground state because an electron may be described as a wave; the ground level is the location where the simplest wave just fits around the orbit. Inside that orbit, a wave will not fit.

Finally, the ball moves from one level to any other equally well. Electrons do not move between all possible energy levels with equal probability. As a result of interactions between multiple electrons, the probabilities differ between certain energy levels and others. For example, for some levels, the probability of a downward jump is low, while other levels behave as if an express elevator existed between the upper and lower energy levels. The express elevator phrase represents a "resonance transition." Some transitions are "forbidden"; the term arises from the early days of atomic spectroscopy. Physicists noted the patterns of the emission lines and connected them to transitions between specific energy levels. They noted that they did not observe other transitions with seemingly similar properties. Using quantum theory, one calculates the rate at which a transition occurs; the "forbidden" transitions have rates close to zero.

Before moving on, let's fold in a bit more terminology. A jump from one energy level to a higher one is called an excitation. When the atom absorbs energy, an electron moves to a higher energy level. That is an excitation because the atom has more energy than it did previously. The reverse process, when an electron moves from a high level to a lower one, is known as de-excitation. An atom that is de-excited emits energy as the electron drops from a high energy level to a low one. Any jump, up or down, goes by the generic name "transition." An electron can also gain enough energy that it leaves the electrical embrace of the atom completely. That process is called ionization; what used to be an atom is now an ion because one or more electrons have been removed, so the former atom now has a positive charge. The "express" jumps described at the end of the previous paragraph are called resonance transitions, because they occur extremely rapidly and with high probability.

The distribution of energy across the electromagnetic spectrum, as carried by the photons we have measured, provides clues to the nature of the astrophysical source. For example, if all the photons from a particular source fall into a given, narrow wavelength region, that distribution of photons tells us something fundamental about the source (regardless of whether we understand what we're being told!).

Arguably the most important step forward, during the early twentieth century, in our understanding of astrophysics is the Hertzsprung-Russell diagram, first assembled by E. Hertzsprung and H. N. Russell in the early 1910s. The diagram plots the intrinsic brightnesses of stars versus their colors. "Color" here means the difference in the measures of brightness in two different bandpasses, generally the "blue" band, centered on 4,400 angstroms, and the "visual" band, centered on 5,500 angstroms. The resulting plot (Fig. 7.1) shows that most stars fall into a thin, nearly diagonal band, illuminating the basic physics: the hotter the star, the brighter. The Hertzsprung-Russell diagram could predict a star's properties.

In the late 1890s and early 1900s, Annie Cannon and her coworkers at Harvard College Observatory laboriously assembled the luminosities and spectra of about five hundred thousand stars to build what, at that time, was the largest catalog of stellar spectra. Cannon and others noted that many stars showed similar spectra. They concocted "spectral types," lumping spectra of similar appearance together into one class. Their approach was the correct one when you do not understand a mass of data: sort and search for similarities or patterns. If a few groups contain most of the behaviors, you can be confident that careful reasoning will find an explanation, even if "careful reasoning" takes years of work.

The Hertzsprung-Russell diagram (Fig. 7.1) appeared at about the same time as physicists hammered out the initial understanding of the structure of the atom. Physicists and astronomers had long noted patterns of emission lines in the spec-

Fig. 7.1. The Hertzsprung-Russell diagram plots a star's luminosity or absolute magnitude (here, the luminosity is represented by the absolute brightness in the green-yellow portion of the optical band) versus its temperature (or color, where the color is defined as the difference in brightness in two difference bands [blue, visual] in the optical). The diagonal band is the main sequence, the locus of points where stars burn hydrogen in their cores. The knot in the lower left are the white dwarfs, stars that are very hot, as indicated by their temperatures, but very small. The Sun lies near the middle of the plot, with a temperature of about 6,500 K and an absolute magnitude of +5. (Plot generated from data obtained from the HIPPARCOS public data archive at the European Space Research and Technology Center of the European Space Agency.)

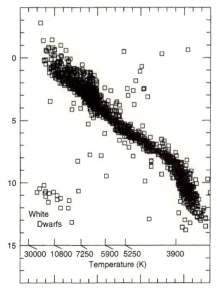

tra of hot gases. What did these patterns signify? What did they explain about the stability of an atom? The Danish physicist Niels Bohr provided the conceptual breakthrough when he pictured an atom as a nucleus with orbiting electrons, a miniature solar system. Others had considered this view but always stumbled over one question: what dictated the locations of the electron orbits?

If electrons orbited the nucleus, then they traveled on curved orbits, which meant they were constantly accelerating. Acceleration is often thought to mean a rapid increase in velocity. "Acceleration" really means a change, any change, in velocity. (The change can be a slow or fast increase, or even a decrease, for which we have the less-used "deceleration.") To the physicist, acceleration means not only a change in speed but also a change in direction. As you pull away from a stop light in your car, you accelerate and change your velocity. If you ride a merry-go-round, you are moving at a constant speed but you are constantly changing direction, so the physicist says you are accelerating. This is not so difficult as it might sound. Velocity is a directed quantity, so it possesses magnitude and direction. A change in either quantity requires an acceleration. "Speed" is the magnitude of the velocity only and is not a directed quantity. Creating directed quantities called "velocity" and "acceleration" then provides terms for the sensation you feel whenever you round a corner at high speed.

Based on the understanding of atoms and radiation that physicists possessed prior to about 1910, accelerating electrons radiated energy. An electron radiating its energy will spiral into the nucleus of the atom; based on these simple ideas, physicists calculated the expected lifetime of an atom: about 10^{-12} seconds. Furthermore, an electron spiraling to the nucleus radiates a continuous spectrum, not the discrete emission lines observed. Physicists faced the following quandary: based on their understanding and theory, atoms could not be stable, let alone emit. But atoms did just that. What made atoms stable? How to explain the discrete spectrum observed? Physics in the late 1800s and early 1900s was in a crisis.

Matter and light can be described either as particles or as waves, a fact established by physics experiments. Those experiments told the physicist that he or she was permitted to use whichever analogy provided the easiest understanding of the data. In the case of atoms, everyone prior to Bohr had used the particle picture. Bohr adopted the wave analogy and showed that the orbits of the electrons could exist only at discrete separations or radii from the nucleus. Why does that solve the problem? Only at specific radii will an integer number of complete wave cycles (peaks plus valleys) fit around the orbit. At other radii, a fractional number of waves is required; fractional waves lead to cancellation of the waves as the peak of one wave meets the valley of another.

The idea of an integer number of waves immediately forced the interpretation that electrons exist only at particular radii. If the electron then gained or lost energy, it had to jump from one orbit to another, either closer to or farther from the nucleus. Such discrete jumps produce emission lines (inward jumps) or absorption lines (outward jumps) in a spectrum. Atomic spectroscopy was born.

For the astronomer, the understanding of the structure of the atom meant that the spectrum of a star or gaseous nebula communicated the chemistry of the star or cloud. We could study the chemical composition of distant objects in the universe from our offices.

How does our model of the atom help us understand emission and absorption lines? Each atom is composed of some number of protons, neutrons, and electrons. Protons and neutrons are located in the nucleus, electrons in orbits about the nucleus. The number of protons determines the type of atom, whether an atom of hydrogen or helium or carbon or uranium. Each element has a unique number of protons (the atomic number).[5] The number of protons plus the number of neutrons determines the atom's mass. The difference in the number of protons and the number of electrons determines the charge. The simplest atom is hydrogen: one proton, zero neutrons, one electron. In atomic units,[6] the mass is one and the charge is zero. (Neutral atoms always have a charge of zero.) Helium has two protons, two neutrons, and two electrons—and so on, for all the atoms in the periodic table of elements. Each element has a relatively distinctive behavior; "relatively" because families of atoms exist (the columns in the periodic table).[7]

The protons and neutrons determine the locations of the orbits or energy levels. Each atom has a unique set of energy levels.[8] Electrons transitioning between energy levels produce a unique set of emission lines (downward transitions) or absorption lines (upward transitions). If we measure the positions of all the lines of an atom and for all possible atoms, then, when we see the same pattern in the spectrum of an astrophysical source, we know the type of atoms present.

If astrophysics were just that simple, everything would have been solved thirty years ago. What complications are there? There are several, but let's restrict ourselves to two.

First, the gas can be moving. When a moving atom emits light, the frequency of the light is Doppler-shifted to either lower or higher energies, depending on the direction the atom is moving relative to the observer. In astrophysical spectra, a Doppler shift changes the frequency (or the wavelength) at which we find the emission (or absorption) line. Although that may sound devastating for the identification of chemical elements, it is not, because the Doppler shift affects the entire spectrum of the source. Each element's pattern of lines changes frequencies. When presented with a spectrum containing emission or absorption lines of unknown elements, astrophysicists usually do not look for a line at a specific wavelength. Instead, they look for a pattern of lines, because the pattern remains constant.

Second, the emitting matter can be composed of gases at different temperatures. An electron can occupy a higher energy level (farther from the nucleus) only if all the lower levels are already occupied or if the electron has been boosted to the higher level by absorbing light of the correct wavelength (excitation).[9] The larger the separation between energy levels, the more energy the electron must absorb to get to that upper level. If the gas is cool, photons will generally not have sufficient energy to boost an electron to an upper energy level. We will not observe any emission or absorption lines corresponding to those upper levels. The gas will only absorb or emit low-energy lines (in other words, lines with long wavelengths).

The previous paragraph focuses solely on photons producing the jumps between energy levels. Another path exists: the collision of two or more "particles." The word "particle" is here in quotation marks only because the word is a placeholder for the names of the colliding objects. Two or more atoms or ions may collide and produce an excitation or de-excitation of one of the atoms or ions. An atom and an electron may collide and achieve the same result; so, too, an ion and an electron, or an atom and an ion. For collisions to be effective at exciting atoms, the number of atoms per volume must be relatively high. If the density is not high, then few collisions will occur and the resulting emission or absorption line will be very weak (perhaps so weak as to be undetectable).

How do we apply this to the stars? The spectra that Cannon and her colleagues had assembled and grouped into spectral classes really showed the influence of the star's temperature on the atoms of gas. Cannon's classes started with spectra that showed few but strong absorption lines and progressed to spectra that showed numerous but weaker absorption lines. The classes were labeled by the alphabet. Once Bohr described his model for the atom, astrophysicists realized that the spectral sequence was a temperature sequence. The presence or absence of particular absorption lines revealed the temperature as certainly as if we dipped a thermometer into the star. Cannon took the original spectral sequence, labeled A, B, C, . . ., combined some letter bins and reordered it as a temperature sequence: the now-familiar stellar spectral classes of O, B, A, F, G, K, and M stars. Stars of spectral class A show spectra of nearly pure hydrogen (see Fig. 6.6, p. 99); above A (O, B), hydrogen is increasingly ionized because the stars are hotter and the hydrogen absorption lines become weaker. Below A, the hydrogen lines weaken because the stars are cooler and the electrons in the hydrogen atom cannot absorb much of the outgoing radiation because it lacks the requisite quantum of energy.

It would take astrophysicists another 50 years to understand much of what the Hertzsprung-Russell diagram tells us, but basic physics could increasingly be used to study the stars in detail. Why 50 years to understand the diagram? How a star responds as it runs short of fuel was not fully understood until the early 1950s. That the diagram is a snapshot of stellar evolution was not appreciated until about that time, either.[10] The details of this diagram required the construction of elaborate numerical models of a star, all of which meant large and fast computers. Sufficiently large and fast computers did not really exist until the 1960s or so.

Let's turn to X rays. X-ray spectroscopy arguably may have its easiest application to the study of the remnants of supernovae. Stars essentially make up all the larger structures we see: star clusters, galaxies, and clusters of galaxies. The evolution of massive stars influences everything: they emit radiation from the radio band to X rays, lose mass to their surroundings, and blow up, perturbing the interstellar medium with shock waves. When they detonate, the explosion spews their inner layers outward. The matter is hot not only for having been inside the star, but also because the explosion itself heats it. Most of the gas in a star is hydrogen, but deep in the interior, the proportion of hydrogen falls while the fraction of other elements rises as the star burns hydrogen. At the core of a star just before it detonates, none of the gas is hydrogen.[11]

The interior of a massive star just before detonation resembles an onion. Slice an onion in half; notice the layers of shells that surround the core. If we could similarly slice a star, it would show layers of gases of differing composition.[12] At the core, iron slowly accumulates; lighter elements fill the overlying shells. At present, we do not really understand the mechanism by which a massive star detonates. We do know that the core must collapse inward because of atomic physics.

A star lives because the outward pressure of radiation and hot gas balances the inward pull of gravity. Throughout a star's life, these forces are in approximate balance; for short intervals, one or the other may become dominant, but the balance is quickly restored. The outward flow of radiation occurs because hydrogen is converted to helium, helium to carbon, carbon to nitrogen and oxygen, and so on up the periodic table. Each reaction is exothermic: it releases energy. The exothermic reactions stop at iron because it is energetically the most stable element:

once formed, energy must be added to break it up (which makes it an endothermic reaction). This is easy to understand. Many people heat their houses with oil, gas, or wood; they do so because the burning of any of these substances gives off heat. If you had to supply heat to the reaction, you would not have a heat supply, but a heat sink—not very useful on a cold winter day.

For a star, the existence of matter at the core that is not supplying energy (not "burnable") to support the overlying layers constitutes a death sentence. As the iron accumulates, it initially resists the pressure of the layers above it simply because it is a very hot gas (many millions of degrees). Eventually, however, too much iron accumulates and the core collapses inward. At this point, a supernova is created.[13] The expanding shock drives the onion layers outward.

Supernovae basically come in two varieties, classified using optical spectra. Type I supernovae do not show emission or absorption lines of hydrogen; Type II supernovae do. Several subclasses exist, but for the purposes here, the broad separation is sufficient. All Type II supernovae are believed to be massive stars; the "onion" description above applies to a Type II supernova. The outermost layers are hydrogen, while inner layers are carbon, oxygen, magnesium, silicon, sulfur, and iron, as well as other elements.

How does all this apply to observations in the X-ray band? The ejected matter is hot; the shape of the spectrum of one supernova, detected by a hard-X-ray/low-energy-gamma-ray detector, required a temperature of about 80 keV, or one billion degrees,[14] to match the observations. Matter at millions or billions of degrees emits X rays, so X-ray studies probe the ejected layers. Emission lines of a particular element emitted by matter in those ejected layers provide at least estimates of the abundance, or quantity, of that element. Accumulating estimates of the abundances of all the elements in the ejected layers help set targets for computer models of the interior of a star. We simulate the interior of a star, including the density, temperature, and abundance of each element present. The computer code explodes the star and calculates changes in the amount of any elements present. We compare the calculated abundance values to the observed values to see if one of the models produces a coherent explanation, coherent in the sense that the model must fit with any other evidence we've managed to obtain. For example, calculated abundances indicating that the progenitor was a low-mass star would not do if we possessed other evidence that the star was massive. Most likely, the computer model would be incorrect.

We make this comparison because we do not know the mass of the progenitor, the star that exploded to produce the remnant under scrutiny. Ideally, we'd know so much about the nature of exploding stars that from a measurement of the amount of an element present in the ejecta, we'd infer the mass of the progenitor. The mass of the progenitor is the key to the doorway of stellar evolution. If we somehow deduce a mass estimate, we then possess a good guide to the steps that led the star to the present moment in its history.

The huge advantage that SN1987A provided astrophysicists was its location in the Large Magellanic Cloud, a neighbor of our galaxy. The Large Magellanic Cloud is a well-studied galaxy because of its proximity. Researchers studying the cloud serendipitously observed the progenitor several years earlier; a spectrum of it even existed (and it looked rather normal for a star about to detonate!). That meant we knew the properties of the progenitor prior to its explosion. Instead of inferring the properties of the progenitor from the debris of the explosion, we possessed data.[15]

To measure the various abundances of the elements, we must use detectors with good spatial and spectral resolution. We need both: with poor spectral resolution, we do not separate the emission lines of each element; with poor spatial resolution, we blend the emission of distinct elements.

Consider an example. From the ASCA image of Cas A (Fig. 7.2), we extract the spectrum shown in Figure 7.3. You can see the spectral resolution is rather good; lines of silicon, sulfur, calcium, and iron are clearly visible. If, however, we look at the image, we cannot see specific emission regions because the ASCA point-spread function was rather poor.

Now compare the Chandra images (Color Figs. 5 and 6 [see insert]). The spectral resolution of the ASCA CCDs, and the resulting spectra, are similar to the Chandra CCD spectral resolution. Chandra's spatial resolution, however, is about a hundred times better (approximately 90 seconds of arc for ASCA versus about 1 second of arc for Chandra). With the higher spatial resolution, we extract from the Chandra data the spectrum of individual filaments (Fig. 7.4). We extract spectra from specific regions, showing the spectra from four different locations in Figure 7.5 (p. 124). Notice how the spectra change from location to location. We can color-code the photons: Color Figure 5 shows an image of Cas A with photons falling at the energies of ~0.9 keV colored red, those at the argon line (about 3.9 keV) colored green, and those at the iron line colored blue. The purple regions (blue plus red) are not visible in the interior but lie near the outer edges. That conclusion is supported by the spectra extracted from the various filaments. The spectrum from the outer eastern filament shows a strong emission line from iron. Earlier, iron supposedly

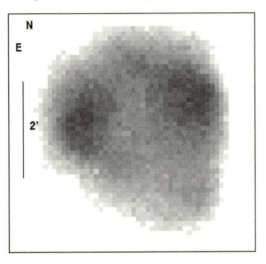

N
E
2'

Fig. 7.2.   The ASCA image of Cas A. From the data that make up this image, the spectrum of Cas A is extracted and presented in the next figure. From previous images of Cas A, the reader should note how few features are visible. (Image generated from data obtained courtesy of the ASCA data archive, NASA-Goddard Space Flight Center.)

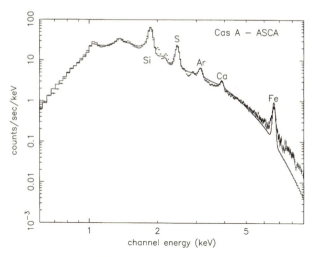

Fig. 7.3. The Cas A spectrum extracted from the data of the previous figure. Note that the emission lines are clearly resolved from the background. Compare these spectra to the descriptions and spectra from chapter 6. The data were obtained by extracting the spectrum from the entire image.

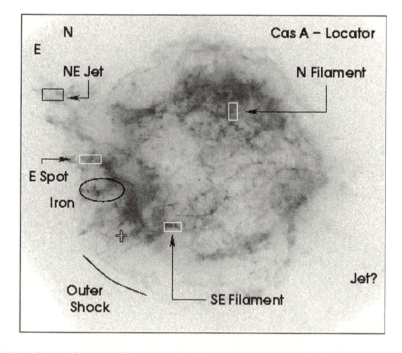

Fig. 7.4. Cas A with regions of interest marked for examination in the color images (circles, lines) and regions from which spectra were extracted (boxes), presented in Fig. 7.5. (Image generated from data obtained from the Chandra public data archive at the Smithsonian Astrophysical Observatory.)

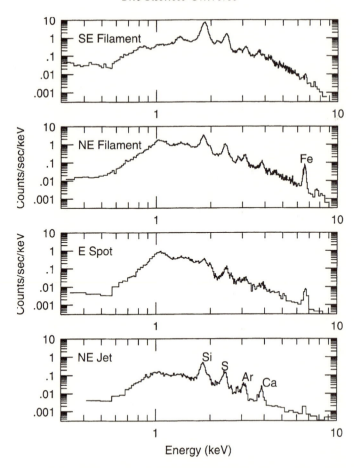

Fig. 7.5.   Spectra extracted from the regions defined in the previous figure. Note the differences in the spectra, particularly between the top two at energies of about 6.5-keV (iron emission), where iron is present in one spectrum and absent in the other. Contrast these spectra with the spectrum in Fig. 7.3. The spectra are quite similar, but the spectrum from ASCA cannot be spatially separated. An ideal X-ray observatory would combine Chandra's spatial resolution, Newton's collecting area (or a larger one), and Astro-E's expected spectral resolution.

accumulated at the core; how did it reach the outermost regions of the remnant? The only explanation requires an overturn of the "onion" structure during the explosion. We know the explosion occurred essentially instantaneously; for the iron to reach nearly the outermost portions of the remnant requires that it, and all of the material originating in the core, move faster than matter from shells lying outside the core. Such a conclusion forms a test: are the outer filaments moving faster? The X-ray spectral resolution is insufficient to measure the difference; Astro-E would have supplied the required spectral resolution. If we alter just one color assignment from the color figure, Cas A appears different. Color Figure 6 (see insert) illustrates this change, in which green shows the events with energies near the silicon line (1.8 keV). These images demonstrate the power of Chandra's spatially resolved spectroscopy, which leads to substantially new science results. Many papers will be written in the coming years from this one image as researchers attempt to match the models to the data and generate new models with refined predictions.

The Cas A image also reveals the limitations of spectral resolution. We cannot separate each element completely because the spectral resolution of the CCD detectors on Chandra and Newton is still insufficient. Ideally, we want to generate a version of Color Figure 6 in which the locations of line emission between the second and first levels of silicon are separated from the locations of line emission between the third and second levels. Such images would provide temperature and density diagnostics for the gas (Fig. 7.6). Our detectors, good as they are, are not yet up to that job.

We could insert a dispersive grating to observe Cas A, but because it is an extended source, the data might be considerably confused. Recall that the grating produces an image of the remnant at the wavelength position of each emission line. The images will overlap for all but a few lines because of the extended size of the remnant. Point sources do not suffer from overlapping images for optics of the quality of Chandra's. To study extended sources with very high spectral reso-

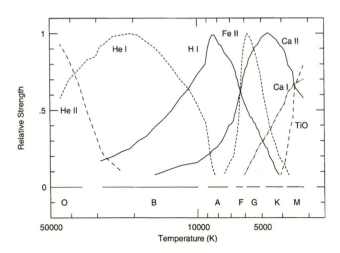

Fig. 7.6. The power of spectroscopy as a diagnostic of temperature. Each curve indicates the strength of the line of the labeled element, ion, or molecule as the temperature of the gas changes. Choose a favorite element and start at a low temperature. As the temperature increases, the strength of the line increases because the temperature of the electrons in the gas producing the line is a better match to the energy levels of the element or ion. Above the peak, the line strength decreases because at higher temperatures, the atoms or molecules are increasingly ionized. Once they are ionized, the locations of the energy levels differ, so the response to temperature changes. Compare the behavior of neutral calcium with that of ionized calcium, or the behavior of neutral helium with that of ionized helium. The symbols are defined as follows: H I = neutral hydrogen; He I = neutral helium; He II = ionized helium; Ca = neutral calcium; Ca II = ionized calcium; Fe II = ionized iron; TiO = titanium oxide.

lution, we need the high image quality of the mirrors of Chandra coupled with a detector that possesses high intrinsic energy resolution—in other words, a detector that does not require dispersal of the light for its spectral resolution. High energy resolution without dispersal of the light is exactly what the Astro E-calorimeter was to provide. The combination of Chandra image quality and calorimeter spectral quality remains a goal of observational X-ray astronomy.

At the highest spectral resolution, X-ray astrophysics starts to provide important data for atomic physics. Several fundamental quantities, such as the lifetime of an electron in an upper level or the rate at which electrons cascade down, are poorly measured, particularly for atoms or ions containing many electrons. This state of affairs is the fault not of the atomic physicists, but of the nature of atomic structure. Certain configurations of atoms are easy to create and to measure in the laboratory. For example, the atom of hydrogen has a simple structure: one electron circling one proton. Its structure is so simple that the atom serves as the basis for teaching undergraduate physics majors about atomic structure because the exact equations can be written down and solved by hand. The next atom in the periodic table, helium, is considerably more complex; two electrons circle a nucleus containing two protons and two neutrons. The complexity of atomic structure increases rapidly beyond helium.

However, if an atom has been nearly completely ionized, so that only one electron circles the nucleus, then the calculation is similar to that for hydrogen, but with a heavier nucleus. Atomic physicists have done a great job in calculating the expected structure of hydrogen-like ions and then measuring their properties in the laboratory to check theoretical predictions. The further a given ion is from this ideal, hydrogen-like state, however, the more difficult the calculations and the harder the laboratory must work to attempt to measure the ion's properties.

X-ray spectroscopy of cosmic plasmas can be considerably easier than laboratory studies. Astrophysicists know that many celestial objects provide the correct temperature and density conditions for particular atoms and ions. In many respects, we only need to locate the right source, then observe and measure. Armed with particular values for density and temperature, laboratory workers can set about re-creating those conditions for a particular gas to check the interpretations. That information then is fed back to the astrophysicist to see whether the original interpretation requires modification. If, for example, the line position is slightly different in the lab than in the celestial source, the gas in the source may be moving; measured motion may alter our understanding of the physics of the object.

Returning to stars, let's answer a question raised earlier. The Hertzsprung-Russell diagram tells us that in the optical band, hot stars are more luminous than cool stars. Surveys of single stars of all spectral types by Einstein and ROSAT showed that cooler stars are bright X-ray sources (spectral types F to K and M), whereas hot stars are relatively weak sources. How can that be? The X-ray spectra of cool stars look similar to an X-ray spectrum from the Sun. Studies of the Sun show that the temperature increases away from the surface. Our intuition says that gas should become cooler the farther it is from a source of energy. In this case, our intuition is incorrect. Some process connected to the Sun supplies the temperature rise with the increase in distance.

The first argument, and one that stood until the mid-1970s, proceeds in the following manner.[16] Stars move heat from their centers to their surfaces in one of two ways—by radiation or by buoyant, hot globs of gas. The temperature at the center of a star is about 10 million degrees (hotter in more-massive stars, slightly cooler in less-massive stars). The temperature at the surface ranges from a few thousand degrees to tens of thousands of degrees. For the more-massive stars, radiation is more efficient at moving the heat from the center to the surface. For the less-massive stars, buoyant globs move the heat more efficiently. The action of moving heat by buoyant globs is called "convection." When you heat water for tea, or to cook pasta, you wait until the water is boiling; the boiling is a sign of convection. On less-massive stars, the surface of the star constantly convects as hot gas from deep within the star rises to the surface.

When a glob reaches the surface, it does so with some velocity. So do bubbles of hot water in the pot on your stove. That's why the surface of the water is no longer flat, but constantly in motion. That motion, at the surface of a star, pushes a shock into the gas above the stellar surface. Rapid observations of the Sun showed the surface in motion as expected. The shock, created by the convecting surface, deposited its energy in the next layer outward. This process is known as "acoustic heating."

The region of a star's atmosphere above the surface is called the chromosphere. In low-mass stars, the chromosphere is a strong source of X rays fueled by convection. Massive stars of spectral types O, B, and A do not undergo convection, so they do not have chromospheres and should not be strong X-ray sources.

Great theory, right? It is too bad that the observations fail to support the theory.[17] Acoustic-heating theory predicts that all stars undergoing convection should have approximately the same luminosity because the convective energy dumped into the stellar atmosphere is roughly identical. Under this reasoning, the Sun should be a relatively bright source. The observations instead show a broad range of luminosities; the Sun falls near the bottom of the observed range. Furthermore, the theory predicts that any star that does not undergo convection will not be an X-ray source. Yet single, isolated O and B stars have been detected in the X-ray band. Finally, observations show variations in the strength of the coronae by a factor of a thousand or more. Acoustic theory did not predict such variations. How does one explain the observed behavior?

The current model postulates that the X-ray emission of single stars of spectral types F to M is driven by the combination of magnetic fields and stellar rotation. The fundamental change in research in the theory of the atmospheres of stars came about by ultraviolet and X-ray observations of the Sun obtained by astronauts on Skylab (the United States' space station of the late 1970s) and from X-ray observations of a broad range of stars using the Einstein Observatory. The details of the magnetic theory still must be worked out, but one key observation provides some support: the rotation rate of the Sun is rather slow, whereas other, more X-ray-luminous stars rotate more quickly. Astronomers have been using and will continue to use the gratings on Chandra and Newton to obtain the data that will help us understand how the outer atmospheres of stars other than the Sun are constructed. These studies tie our detailed observations and increased understanding of the Sun into the wider range of stellar properties.[18]

Fundamentally, everything we know about the universe has come from electrons and atoms or ions. If electrons absorb light, we see an absorption line. If they emit light, we see an emission line. If they crowd together, we see a continuum. If they move in bulk, we see Doppler shifts. By studying the levels of various atoms and ions, we learn the conditions under which specific transitions occur. This knowledge gives us the ability to start interpreting the spectra we collect.

8

"If you have an important point to make, don't try to be subtle or clever. Use a pile driver ..."

—Winston Churchill

Consider this chapter a summary of a theme that wove its way through previous chapters: why is the study of the X-ray universe so important? We have learned how to measure the position, arrival time, and energy of an X ray. We have learned that X-ray-emission lines report information similar to that provided by optical lines, although we do not yet have sufficient spectral resolution or sensitivity to extract all of the available science. Nothing describes the goals of science better than one of the grand questions adopted by NASA as part of its road map: why does nature repeatedly choose similar solutions to particular problems? This question lies at the core of science: the identification and understanding of the patterns evident in the universe.

Return for a moment to Seurat's *Grande Jatte*. Earlier, we treated this painting as an analogy to understand the point-spread function. Let's use the painting again, this time to understand something fundamental about the universe. X rays are produced by energetic processes. The amount of energy required to create a typical, 1-keV X ray could produce about a thousand optical photons. Relative to the number of optical photons, there are very few X-ray photons. Each X-ray photon carries about a thousand times as much energy,[1] but there are many fewer photons to be collected. This behavior has nothing to do with our instruments, but rather is an observable property of the universe in which we live. This is why X-ray images look grainy: the images essentially show each detected X ray; only with the high-quality mirrors of Chandra, with their high spatial resolution, do X-ray images approach the continuous quality of optical images. Imagine *Grande Jatte* as both an optical and an X-ray image and assume that whatever process creates photons does so with equal, total energies. Remove about 999 of each 1,000 dots of color in Seurat's painting. What is left will be a crude analogy of a typical image of an astronomical X-ray source.

If X rays are more energetic than optical photons and are produced by hot gas, how hot must the gas be? The answer is, at least a thousand times hotter. If the efficiency with which X-rays are produced is factored in, the temperature must be still higher. The Sun has a surface temperature of approximately 5,800 degrees Kelvin; the outgoing thermal radiation from the surface peaks near the yellow-green portion of the spectrum and is approximately equal in the blue and red bands of the optical spectrum, leaving the Sun with an overall yellow color.[2] If the Sun's surface were more than a thousand times as hot, the surface would emit X rays. But the Sun is an X-ray source, even if it is a weak one. How then does it produce any X rays? How do other sources produce X rays?

These questions form the basis of a description of an observational astronomer's research: infer the physics of the process's or processes' contribution to the source's emission; do so with support solely from the emitted light detected.

Let us return to the question of how the Sun produces X rays. There are basically two paths: thermal processes and (not to be too clever) nonthermal processes. The distinction depends on how the electrons produce and radiate their energy.

We know that "thermal" means heat, so thermal processes require hot gas. A gas (or a liquid or a solid) stores its heat in the energy of its moving electrons and ions. Electrons in motion in a hot gas move quickly. If an electron collides with another atom or molecule, or with a photon, or is deflected from its trajectory because of the presence of a nearby charged particle, it surrenders or radiates some or all of its energy. In a collision with an atom or molecule, the impact of an electron creates an ion by ejecting another electron. An atom or molecule may absorb the ejected electron, creating an excited state. An electron may stimulate an excited atom or molecule to release its excess energy. Some of these transitions will create X rays if the difference in energy between the starting and ending energy levels is sufficiently large. A free electron, deflected from its trajectory by a nearby ion, emits radiation. The scientific term is *bremsstrahlung*, which is German for "braking radiation." The collision of an electron with a photon alters the energy of both; if the electron gains energy from the collision (the photon initially has more energy than the electron), the process is called "Compton scattering";[3] if the electron loses energy to the photon, the process is "inverse Compton scattering."

The preceding paragraph describes how fast-moving electrons produce X rays. How are electrons accelerated to the necessary high speeds? In an astrophysical setting, hot gas is produced by the interaction of a shock wave with gas or by accretion processes.

A shock wave is caused by a discontinuous change in the pressure, temperature, and density of a gas. Let's sketch a picture of the events (but not the physics). Imagine walking through the lobby of a building that has walls of glass. The glass has just been cleaned, so it is extremely clear. You walk directly into one of the glass panels. What happens? Your forward motion, as measured by the speed of your walk, changes instantaneously; in this case, to a dead stop. Portions of your body such as your nose deform upon encountering the glass. Your dignity no doubt undergoes a profound change. What the glass has done to you is a crude picture of what a shock wave does to an atom, with one difference.

The atom, or cloud of gas, suffers the reverse situation: whereas the glass dropped your speed to zero, a shock wave gives the atoms in the gas cloud a kick. Shock waves start as pressure waves from an explosion, for example, and are transformed into shock waves as they propagate. A pressure wave has a wavelength, but its amplitude is not a smooth ripple up and down. Instead, the pressure wave has a region of compression and a region of rarefaction (think "back and forth"). The concept of compression and rarefaction stands out in traffic flow. When cars are packed together on the roadway, for whatever reason, they create a compression region; the leading cars speeding away define the rarefaction region.

Pressure waves move at the speed of sound in the medium through which they propagate. As such, the medium fundamentally affects the wave. The speed of sound in a medium increases if the medium's density is higher and decreases if the density is lower. As the pressure wave propagates through matter, the region of compression speeds up, but the region of rarefaction slows. The compression part of the wave catches up, distorting the wave front from a slowly varying wave to an abruptly changing one.

Everyone who has been to the ocean has observed one part of a wave catch up to another part. Ocean waves approach the shoreline as moving ripples or mounds while they are far from shore and in water more than about five times their height. The leading edge of the wave slows as it moves into increasingly shallower water. The trailing edge of the wave catches up, forcing the wave to increase in height. The wave steepens and ultimately breaks. Please note that an ocean wave does not become a shock wave just because the wave shape distorts. Ocean waves are displacement waves with the wave-front distortion represented by a change in height. Shocks waves from an explosion are pressure waves and do not have height. The explanations for the distortions of the waves differ between the two cases, or, to put it differently, the physics of the wave changes, differ. Use the picture just described with care. As an aid to illustrate how a wave distorts, the picture is useful.

Explosions, such as supernovae, create shock waves;[4] the shocks propagate through the matter that formed at one time the outer layers of the star and through both the matter lost from the star during its evolution and the interstellar medium. Shocks have another property: they reflect. An outward-propagating shock gradually sweeps up matter in its path. This slows the shock and increases the pressure; when the pressure is sufficiently high, a reverse shock is driven back into the already-shocked matter. Essentially, the outgoing shock reflects an inward-moving shock. Most supernovae are believed to emit X rays by this outgoing/reverse mechanism. Color Figure 7 (see insert) shows the radio, optical, and X-ray emission from the supernova remnant E0102-72, a remnant in the Small Magellanic Cloud, another neighbor of our galaxy. This one image essentially defines the textbook example of shock physics. E0102-72 appears ringlike because the expansion is confined largely to a plane.[5] The radio emission comes primarily from the outgoing shock wave; the X-ray emission comes mostly from the inward-moving, reflected shock wave. The optical filaments arise from the nature of the explosion. View the explosion as a ram pushing against the outer layers. The ram imparts its momentum to those layers, driving them outward. Replace the ram with gas of a density higher than the gas in the outer layers. The high-density gas pushes outward against the low-density gas. The outgoing shock pushes on the outer layers creating fingers that poke into both layers. This situation is equivalent to that of a dense liquid placed over a light liquid in a gravity field (an experiment you can try at home).[6] In E0102-72, this mechanism, known as the Rayleigh-Taylor instability, likely created the oxygen filaments (see Fig. 8.6, p. 141).

How can we tell if a shock is present if we do not have X-ray data? If the shock is sufficiently hot, it produces energetic X rays; nearby atoms, even if "nearby" means a significant fraction of a light year apart, may absorb some of the X rays. The X-ray photons ionize the atoms, a process called "photoionization." If the ion remains in the same region, additional ionizations may occur until the original atom becomes highly ionized.[7] The thermal radiation emitted by atoms is usually characteristic of the temperature of the gas. A photoionized atom, however, behaves as if the gas had a considerably higher temperature; the higher the ionization state of the atom, the higher the temperature of the gas. For example, optical emission lines from photoionized atoms always look out of place with respect to the temperatures and densities implied by the other emission or absorption lines present in the spectrum. They literally stick out like that one unmowed blade of grass. Emission lines of nine-times-ionized iron are occasionally detected in the optical spectra of supernovae. To remove nine electrons from iron requires gas temperatures of about a million degrees. Shocks easily reach this temperature. The presence of the emission line suggests that a shock may be present, although other explanations are also possible. X-ray spectra provide a direct assessment of the conditions, but occasionally, we see the fingerprints in other bands of the effects of X rays.

Accretion is the process by which material is accumulated at a specific location. To build a snowman on your front lawn, you start rolling a small ball of snow. As you roll it, the ball accumulates, or accretes, more snow until it is too large for you to move. In astrophysics, accretion occurs because mass attracts more mass. The smaller mass accelerates toward the larger mass. Constant acceleration means that the speed of the falling mass increases. When it reaches the attracting mass, it collides with it and transfers its accumulated energy to other particles, so all move faster. Accretion processes heat gas. When the gas temperature or impact energy becomes sufficiently high, X-ray production becomes favorable.

Accretion occurs essentially by one of two paths: either through a disk or quasi-radially, which means matter falls inward toward the accretor from all or nearly all directions. Radial accretion occurs in isolated neutron stars if the neutron star passes through or near a cloud of gas or accretes solely from the matter randomly distributed between the stars (the interstellar medium). Quasi-radial accretion may occur in a binary system provided one of the stars loses mass via a strong stellar wind. For these binaries, matter does not transfer from one star to the other by a Roche-lobe overflow process. Instead, a normal star, lying within its Roche lobe, loses mass through its wind. If its companion, a compact star such as a neutron star, accretes as it orbits, it will be bathed by the wind. The details are complex because the compact star behaves like an obstacle to the outflowing wind. Theoretical simulations demonstrate that a bow shock forms on the leading side of the compact object as it plows through the stellar wind.

How do we know that accretion processes generate X rays? Several reasons exist. Theoretically, a few lines of calculations show that accretion onto a compact source easily generates X-ray emission.[8] Second, the efficiency of the process is high. Third, the transfer of mass from one star to the other in a binary also transfers angular momentum. We expect the accreting object to spin faster as it accretes matter from its companion. We observe spin-up in neutron-star binaries; accretion provides the simplest explanation.

The nonthermal process that generates X rays requires the presence of a magnetic field. In a magnetic field, electrons spiral about the field lines. If the field is sufficiently strong, the spiraling occurs with a short radius, so the acceleration is high. Recall that acceleration is change in speed and change in direction. For electrons spiraling around magnetic fields, the "change in direction" part of the definition is particularly apt. Accelerating, free electrons (free of any atom) radiate and give off "synchrotron radiation."[9] Let's look briefly at two extremes of magnetic field strength.

The Sun has a magnetic field, but it is relatively weak.[10] The overall solar magnetic field is about 10 gauss (a unit of magnetic-field strength; the magnetic field of the earth is 1 gauss). Still, it influences the dynamics of matter in the solar system, driving a flow of charged particles outward (the solar "wind"). The eruption of a magnetic-field line through the solar surface creates a sunspot. The spot is the portion of the process visible to ground-based telescopes. The spot appears dark only because of the surrounding bright, hot gas. Skylab X-ray images showed the existence of "coronal holes," regions of little or no X-ray emission. In these regions, the magnetic field of the Sun streams directly outward. The aurorae (northern or southern lights) are produced when charged particles from the Sun, funneled to the polar regions by Earth's magnetic field, hit Earth's atmosphere.

Turn up the magnetic field by a factor of a thousand trillion (about 10^{15}).[11] Electrons respond, spiraling tightly around the field lines. The Sun is a weak X-ray source, overwhelming sensitive X-ray detectors such as Chandra's CCD spectrometer only because of its proximity. The strongest steady field generated in a laboratory is 4.5 mega-gauss; the strongest pulsed field, "pulsed" meaning that it exists for a very short time, is 85 mega-gauss (8.5×10^7 gauss).[12] Moving in a trillion- to peta-gauss field, electrons generate X rays and gamma rays. This is the world of the "soft-gamma-ray repeaters," objects that produce bursts of X rays and gamma rays.

Soft-gamma-ray repeaters, or SGRs, have only recently been recognized as a class of astronomical object. The first object detected occurred on March 5, 1979, when a gamma-ray burst lit up all the satellites in the solar system that carried onboard gamma-ray detectors. The abundance of detectors permitted scientists to triangulate the source of the emission; they narrowed the positional uncertainty to a small region of space in the Large Magellanic Cloud. In that direction existed a known young supernova remnant, so the association tantalized astronomers. The source burst 16 times over the next four years.

Additional observations through the 1980s led to the recognition of SGRs as a class. Because they undergo gamma-ray bursts, astronomers initially lumped them together with the other, normal gamma-ray bursts. In contrast to normal bursts, SGRs exhibit soft-gamma-ray spectra and they recur. Currently, four are known;[13] a fifth awaits confirmation. Three are associated with supernova remnants. The current theory invokes a neutron star with a magnetic field of a trillion gauss or more. The bursts are believed to be generated by a star quake in which the surface of the neutron star is disrupted slightly. The density of matter at the surface of a neutron star is so high that the resulting seismic wave releases a huge burst of energy.

The star-quake model recently received a boost of support when observations of an SGR showed that it had recently "glitched." Occasionally, neutron stars exhibit "glitches" (that *is* the technical term), or sudden changes, in their times of arrival of the pulses. A rotating neutron star emits a beam of energy (for reasons not yet fully understood) much like the beam from a lighthouse. If Earth happens to lie in the line of sight of that beam, we detect pulses from the object as it rotates. The pulsar in the Crab Nebula rotates about 33 times per second; high-speed imaging or photometry detects each pulse. Measuring the arrival times of pulses is relatively easy. High precision is readily achieved and the measurement to eight or more digits in the period between pulse arrival times is routine. With such precision, changes in pulse arrival times stand out. Arrival times for rotating neutron stars gradually separate because the emission of energy slowly robs the neutron star of its spin.

On a plot of pulse period versus calendar time, the gradual lengthening of the period shows up as a steady, diagonal line. The slope of the line is a measure of the rate at which the pulsar slows. A glitch is a sudden change in the pulse arrival time and reveals itself when the diagonal line breaks and jumps to a slightly different period before resuming the steady upward increase. The gradual spin-down and the size of the glitch provide an estimate of the strength of the magnetic field. For the SGRs, the magnetic field strength lies between 200 trillion and 800 trillion gauss.[14] With such a strong field, the spiraling electrons easily generate high-energy photons, including gamma rays. What causes the glitch? One scenario suggests that a small crack occurs in the outer layers, as in an earthquake. With such a high magnetic field and spinning as rapidly as measured, any small change releases large amounts of energy. The detailed understanding remains a topic of research, largely because we do not understand the equation of state of matter at the densities typical of neutron stars.

Consider one last pattern that nature repeatedly uses to solve particular problems. We've discussed shocks, accretion disks, and magnetic fields. Jets are a fourth. A jet is a highly collimated, or focused, stream of fluid or gas. Astronomers study jets across the electromagnetic spectrum. They appear to be intimately connected to accretion processes and magnetic fields; their origins lie in the inner accretion disk.

Astronomers observe jets in young stellar objects and active galaxies. In young stellar objects, jets stretch across large regions of the sky, connecting seemingly disparate phenomena. In active galaxies, they expel material across hundreds or thousands of light-years. Usually only one jet is visible in an active galaxy. Material in the jet of an active galaxy moves at a substantial fraction of the speed of light; such a jet is called "relativistic." Relativistic jets do not radiate in all directions. Instead, their emission is beamed, or confined to a narrow cone, along the direction of motion. We see only the jet that is pointed in our direction; the other jet, pointed away from us, is invisible because it does not emit radiation in our direction. This behavior is typical of relativistic jets; the jet in the active galaxy Centaurus A (NGC 5128 [see Figs. 8.1 to 8.3 on next page or Fig. 2.11 in chapter 2 for the active galaxy PKS 0637-75]) is a prime example. Jets usually terminate in a cloud of material that receives the beamed energy. A physicist, familiar with particle accelerators, would describe the cloud as a "beam dump," the place where the beam's energy dissipates when the research is completed. Astrono-

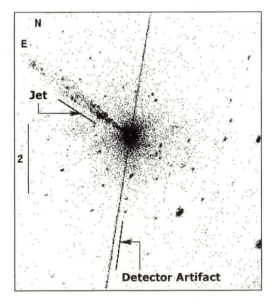

Fig. 8.1.   Centaurus A, an active galaxy in the X-ray band as observed by Chandra. In the X-ray band, the nucleus of the galaxy appears to be nothing more than a jet. (Image used by permission of Dr. Ralph Kraft, Smithsonian Astrophysical Observatory; the image originally appeared in *Astrophysical Journal* 531 [2000]: L9.)

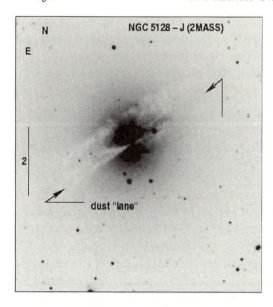

Fig. 8.2. A near-infrared image of Centaurus A. The galaxy is also known as NGC 5128. It is famous for the thick lane of dust (the white band) that obscures the nucleus of the galaxy. (Image generated from data obtained from the public 2MASS data archive at the Infrared Processing and Analysis Center, California Institute of Technology.)

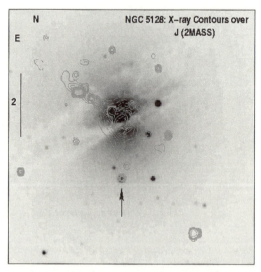

Fig. 8.3. An overlay of the X-ray and near-infrared images of Centaurus A: the jet protrudes directly from the center of the galaxy. (Image generated by combining the two previous figures.)

mers essentially use the beam dump to infer the properties of the beam. The clouds radiate in all directions; one clue to the presence of a relativistic jet in some active galaxies is the existence of two clouds of gas but only one jet.

Jets appear to be produced by accretion disks containing magnetic fields. A rotating magnetic field winds up, becoming a helix that collimates, or focuses, an outflow of matter. This does not mean that jets can be produced only by accretion disks; other methods exist. However, magnetized accretion disks appear to create collimated outflows rather easily.[15] The combination of jets and accretion disks creates a difficult problem for study. Fundamentally, simulations of accretion disks require a minimum of two dimensions; if a magnetic field is added, the problem now requires three dimensions. Furthermore, because a magnetic flow exists, the equations must include magnetic forces as well as fluid-dynamic ones; the resulting problem is one of magneto-hydrodynamics, a mouthful if there ever was one. Jets are an area of active research, and X-ray observations probe regions close to the spot where they form.

X rays reveal the existence of energetic processes. Astrophysicists had not dreamed of some types of binary stars prior to their detection by X-ray observatories. Astrophysicists discovered the strongly magnetic interacting binary systems, known as "AM Her stars" after the first binary investigated, to be strong X-ray emitters in the 1970s. The AM Her stars are also known as polars (polarized stars) following the detection of AM Her as a source of polarized light.[16] Strong magnetic fields emit polarized light because the electrons, responsible for absorption and emission, vibrate or circulate in a preferred direction under the influence of a magnetic field. The reader can see the effects of polarization relatively easily using two pairs of polarized sunglasses. Light reflected from a horizontal surface at a shallow angle becomes polarized. The electromagnetic wave has an electric field component that vibrates vertically, attempting to push the electrons in the surface material up and out and down and into the reflecting surface. The electrons cannot vibrate in those directions. The wave also contains a horizontal component, causing the electrons of the reflecting surface to vibrate back and forth in the surface. This motion is possible. As a result, the reflecting surface suppresses the vertical component of the wave and reflects the horizontal component. Polarized sunglasses essentially have the molecular equivalent of horizontal slats. Vertically polarized light passes through, but the slats stop horizontally polarized light. Glare is the shallow-angle reflection of light off car windshields, chrome, painted surfaces, and the like. Polarized sunglasses eliminate that light. To see the effect, take two pairs of sunglasses, hold them up to the light, and rotate one pair so that it is 90 degrees to the other pair. The rotated horizontal slats are now vertical, so light attempting to pass through both pairs of lenses will be blocked completely.

Recall the interacting-binary picture that includes a mass-transfer stream and an accretion disk surrounding a white dwarf. For AM Her stars, the accretion disk does not exist because the magnetic field of the white dwarf is so strong that it prevents the disk's formation. Mass from the secondary essentially threads onto the magnetic-field lines immediately after leaving the vicinity of the secondary. Accretion occurs not near the equator, as in a binary with an accretion disk, but near the magnetic poles. The current model for these systems envisions something similar to the aurorae of Earth, but orders of magnitude hotter and denser as the mass rains down on the magnetic polar region.

At heart, nearly every astronomer is a closet cosmologist; we all want to figure out how the entire universe evolves and how it originated. The investigation of the evolution of the universe is a goal for X-ray astronomy, too.

The physics of stars, and the knowledge of how they evolve, became understandable only because of spectroscopy and quantum physics. Eliminate both, and about 90 years of research on the interior structure of stars, the evolution of stars, and how stars are born and die, vanishes, as well as our knowledge of the production of the chemical elements. Everything we touch is composed of a large fraction of the 92 naturally occurring chemical elements that make up the periodic table. Our understanding of the origin of those elements is arguably one of the most stunning revelations of science. Copernicus moved Earth from the center of the solar system. Harlow Shapley and his colleagues, in the first third of the twentieth century, moved the solar system from the center of the galaxy outward toward its edge. Edwin Hubble found an expanding universe. But in the late middle of the twentieth century, as astrophysicists such as Martin Schwarzschild of Princeton University worked out the basic evolution of stars, researchers realized that stars are the only sites in the universe where elements heavier than helium could be made.[17] If a star explodes or blows off its outer layers, those elements escape the gravitational field and become the air we breathe and the iron we forge. While Copernicus and those who followed moved humankind further and further from center court, stellar astrophysicists revealed the intimacy with which we connect to the universe.

The clue to understanding how X-ray astronomers investigate cosmic evolution lies with the supernova remnant Cas A. When a supernova explodes, it ejects nucleosynthesized material into the interstellar medium. Stars forming from molecular clouds contaminated by the synthesized material then begin life with a higher fraction of heavy elements. Once they die, they eject their waste products, further contaminating the next cycle of stellar birth. Imagine a galaxy containing several versions of Cas A. Run the image forward in time and watch as old Cas As fade and new ones appear. Gradually, the nucleochemistry of the galaxy shifts. Place the galaxy into a cluster from which none of the hot gas escapes. Evolve for several billion years.

Each of the galaxies in a cluster emits a considerable amount of energy in the optical band. You could spend a lifetime studying all of the galaxies that compose a given cluster. However, by doing so, you would miss the forest for the trees. The X-ray emission of a cluster of galaxies shows the gravity well of the entire cluster. Essentially, "clusters of galaxies" is a poor term for these objects; properly, they should be called something like "hot gas spheroids with contaminating specks," the galaxies being the specks.[18] The cluster's mass holds the hot gas to the cluster. Clusters of galaxies are essentially the largest structures that are held together by their own gravity. Most of the visible mass in the universe sits in the gravity wells of clusters of galaxies as hot, X-ray-emitting gas. The mass tied up in the hot gas is about a hundred times more than the sum of the masses of the individual galaxies in the cluster.

In the past few years, we have learned that the universe is accelerating; the evidence for this is not yet conclusive but is strong. Experiments to measure the variations in the microwave background radiation establish that the density of the universe is, within the errors, equal to the critical density at which the geometry of the universe is flat. We sum the mass we see, whether it is visibly or gravitationally detected. The sum yields a critical density of about 0.33. Two-thirds of the universe must be a form of energy of which we have no knowledge at present. Of the approximately one-third we detect, a sizable fraction of it remains undetected by any telescope; this is the dark matter. About 4 to 5 percent of the universe appears in a form of matter familiar to physicists. Most of that mass, however, exists as hot gas in clusters.

If we study clusters of galaxies as a function of cosmic time, we should detect the overall evolution of the clusters themselves. If clusters formed when the universe was young, we expect little detectable evolution because sufficient time elapsed to pull into the cluster's gravity well all nearby objects. If clusters formed recently, we should see additional clusters, or at least additional structure, as we look back in time. Why the focus on clusters? Because they are physically large, they are easily detected even at large redshifts.

Clusters are also believed to be relatively simple structures, so the data are easily analyzed. As with so much else in nature, when one looks more closely, the story often becomes complicated. Chandra observations of clusters show a considerable variety among the supposedly simple balls of gas. Our introduction to clusters came by way of Abell 3667 earlier, but with the Chandra image was deferred. Figure 8.4 shows the Chandra image of A3667. Note the sharp edge in the image, not seen previously because the spatial resolution did not exist. The origi-

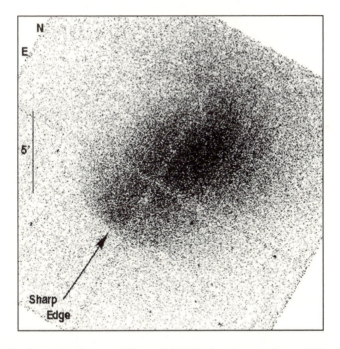

Fig. 8.4.   The Chandra X-ray image of cluster Abell 3667, first shown in chapter 2. Were this image from ROSAT data, it would appear similar, except the sharp edge would be fuzzy. The sharp edge is a new result from Chandra: extracting a spectrum from these data shows that the edge is not hot, so it is not a shock. The edge has instead been interpreted as the equivalent of a cold front as cool gas from one cluster moves into the hot gas of a second cluster. The sharp edge delineates the front of the infalling gas. (Image used by permission of Dr. Alexey Vikhlinin, Harvard-Smithsonian Center for Astrophysics; the original image was presented in A. Vikhlinin et al., *Astrophysical Journal* 549 [2001]: L47.)

nal proposal to observe this object argued that A3667 is an example of a merging cluster, one in which two smaller clusters are becoming one. The proposers suggested that a Chandra observation would detect shocks, evidence of a high-speed collision. The sharp edge, however, is not a shock; it is a pressure wave, indicating that the colliding clusters merge subsonically—in other words, at less than the speed of sound—so no shock forms. Other clusters do show X-ray shocks. Several clusters show holes or gaps in the X-ray emission; the gaps align with bubbles of radio emission. The anticorrelation suggests that magnetic fields contain the radio gas or at least reduce the brightness of the X-ray gas so that it appears as a hole.[19] Will there be a progression in appearance with distance? The rate of formation of the structure depends sensitively on the mass density, the amount of mass per volume, of the universe. If nearby clusters show considerable substructure and evidence of ongoing mergers, then presumably clusters formed recently. The detected fronts and shocks are visible only in the X-ray band. Whatever the answer, it should be found in a few years.

Evolution also occurs in the chemical elements, from relatively pristine material, contaminated by few generations of stars, at small values of cosmic time (high redshift) to recent (nearby) clusters that contain gases cycled through many generations of stellar birth and death. In spite of the new complications offered by the high spatial resolution of Chandra, the hot gas in clusters remains relatively isothermal; the gas is nearly the same temperature regardless of its location in the cluster. Observations of the chemical evolution remain one of the goals of X-ray astrophysics of clusters. The hot gas in a cluster essentially contains the nucleosynthetic history to that time; by looking back at clusters of many ages, we read the trend. Nucleochemical evolution provides an independent look at the evolution of the universe.

Without X rays, we do not see the locations of hot gas. Nothing illustrates that point better than the electromagnetic emission from clusters of galaxies and supernova remnants.

Figs. 8.5 through 8.7 show three images, at different wavelengths, of the supernova remnant Cas A, an object threading its way throughout this book. The radio and X-ray images appear quite similar. Look closely at the optical image. The presence of a remnant is difficult to demonstrate; only a few wisps of gas are visible. Think of the amount of energy flowing outward in each band. If a scientist wants to understand Cas A, optical wavelengths may not be the best band in which to work because the emission from the full remnant is comparatively small. That does not mean information is unavailable; on the contrary, the optical emission lines provide estimates of the densities and temperatures of the filaments. As Figure 8.6 shows, however, not all filaments are equally visible. The original statement stands: if someone forced you to choose one wavelength band with which to study Cas A, the optical band would arguably not be the best choice.

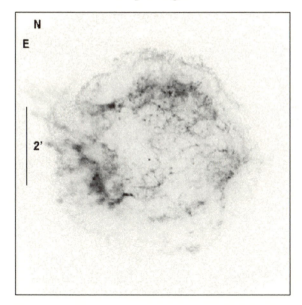

Fig. 8.5. Cas A as observed in the X-ray band by Chandra. (Image generated from data obtained from the Chandra public data archive at the Smithsonian Astrophysical Observatory.)

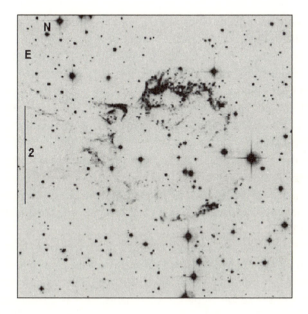

Fig. 8.6. Cas A as observed in the optical. Note the relative lack of emission. (Image used by permission of Dr. Robert Fesen, Dartmouth College.)

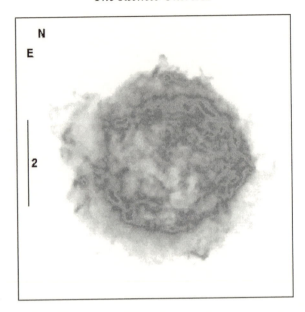

Fig. 8.7.   Cas A as observed in the radio band using the Very Large Array (VLA) of the National Radio Astronomy Observatory. The spatial resolution of the X-ray and optical images is similar to that of the VLA image. (Image obtained from the public CD-ROM "Images from the Radio Universe" [1992] distributed by the National Radio Astronomy Observatory; the original image appeared in M. Anderson et al., *Astrophysical Journal* 373 [1991]: 146.)

Both remnants and clusters illustrate the need to consider all of a source's emission when attempting to sort out the nature of astrophysical sources. For this reason, astronomers, often historically self-described as "X-ray astronomers" or "radio astronomers," increasingly use whatever data are available, regardless of bandpass, to understand the object of study. Astrophysics is the study of the universe, largely carried out by inferring the nature of objects based on the light we collect.[20] Increasingly, we need the light from as many bandpasses as possible to understand the science.

X rays, because they are generated only at very high temperatures, are usually found near the center of the action. The hottest gases generally occur where the physical processes are the most violent, the most dynamic, or the most energetic. X rays are therefore important precisely because they provide a means to study those dynamic and energetic processes. Without X rays, we would have to infer the nature of the processes without the benefit of a direct view. That situation is equivalent to understanding a baseball game by focusing only on the outfielders. Clearly, the fan gets a decidedly narrow view of the game by watching everything but the center of the action. Supernova remnants, such as Cas A, and clusters of galaxies provide examples of that situation.

9

"Destiny is no matter of chance.
It is a matter of choice . . ."

—William Jennings Bryan

Three X-ray satellites: Chandra, Newton, Astro-E. Why so many? Why three for the X-ray band when only one optical telescope exists (the Hubble Space Telescope)? How did NASA propose to Congress to build Chandra? Who chooses the satellites to be built and launched? How is observing time allocated? All of these questions involve the concepts of scientific merit and scientific return.

Scientific return is like beauty, a slippery concept when you try to define it. Most scientists can point to examples of good, and bad, scientific return. Perhaps you think that scientific return may be determined simply by "How many questions did the project answer?" That approach suffers because it is difficult to decide when an answer exists to a scientific question, given that an answer usually generates more questions. Scientific questions are all too often not answered directly; instead, they force the scientists to refine their original questions. This is the fundamental nature of science: a seemingly endless refinement of questions that leads to new investigations focused on the increasingly specific. This nature explains both why science is so successful and why people can be frustrated in their attempts to obtain a straight answer from a scientist.[1] We do not always have straight answers to give.

Scientific answers always come with asterisks, caveats, and other hedges because the practitioners understand that the answer may not ever exist and certainly is not known yet. This point is probably the single most difficult concept to convey to nonscientists.

Our understanding is constantly evolving as we carry out additional experiments or make further observations. A particular question leads to an experiment that generates more questions, which lead to other experiments. Some of those experiments will return results that will be counter to the predictions. Which is wrong, the experiment or one of the assumptions of the theory? Progress in science is not an air-zipping arrow piercing the target. It wallows and becomes stuck in the mud and flashes across the veldt like a cheetah after a dik-dik and chases its tail and runs into a brick wall. That is on a good day, and only if theory and observations exist.

For a science to be healthy, a dynamic balance must exist between theory and observations. Without observations, scientists test little of the theory, so the branch of study withers; without a theoretical framework in which to place observations, specific questions cannot be answered. The field of study degenerates into a collection of data devoid of interpretation. The paths of research then wander aimlessly.

When a dynamic balance exists, theory will explain a set of observations and it will make predictions for additional, more-detailed observations. Once a scientist carries out those detailed observations, the results will often push the theory to predict more precisely. Rapid progress occurs as observers challenge theorists to predict more details and theorists challenge observers to collect higher-quality data.

For example, Einstein's general theory of relativity, published in 1915–16, describes the physics of gravitational fields. Einstein himself verified one of the theory's predictions in 1916 (the detailed behavior of the orbit of Mercury). In 1921, the British astronomer Sir Arthur Eddington observed the deflection of light by a gravitational field during a solar eclipse, verifying a second prediction of the theory. The theory possessed a rather firm observational foundation at that point. Little work occurred in gravitational physics for another 40 to 50 years because specific astrophysical observations did not exist that required general relativity. The discovery of quasars and the realization of their extreme distance provided one of the motivating observations for the resurgence of general relativity.

A more recent example might be more familiar. Physicists have studied superconductors, materials that show little resistance to the conduction of electricity, since H. K. Onnes first demonstrated the behavior in 1911. A theory of how metals become superconductors was published in 1957 by J. Bardeen, L. Cooper, and J. Schrieffer. It explained why the materials superconducted, and it predicted properties that were subsequently verified in laboratory tests. The field appeared to be dormant by the mid-1980s because tests on known superconductors matched the theory almost too well, suggesting that there was little more to learn. In addition, the materials all superconducted only at very low temperatures (a few degrees Kelvin), temperatures that require liquid helium as a cooling agent. Liquid helium is expensive, so few practical benefits of superconductors developed. Researchers moved into other lines of inquiry. That changed in 1986 when Georg Bednorz and Alex Muller of IBM-Zurich discovered superconductivity in a metallic ceramic, a behavior not predicted by the theory. In addition, the temperature at which the material superconducted was among the highest values ever recorded. In short order, other, similar materials with still higher superconductor temperatures were found. These discoveries revitalized the field essentially overnight.[2]

In astrophysics, observations often outpaced theory because, while students wrote the basic theory on a sheet of paper, the actual calculations required were too difficult to carry out without the use of a computer. Many sophomores or juniors in a college-level introductory astrophysics course learn the basic equations that describe the interior structure of a star. Astronomers worked out the basic theory in the 1920s and 1930s and developed our understanding of the source of energy for stars from the late 1930s through the early 1950s. Detailed solutions of those equations, however, did not really start to appear until the 1950s and 1960s. To compute the interior structure of a star, one essentially divides the star into a series of thin layers. The computer then calculates the temperature, pressure, density, and energy flow for each layer, taking care to handle the flow of radiation at many wavelengths. This is very repetitious; computers are excellent tools for just such simulations, and they have restored the dynamic balance in astrophysics, as they have in many other fields of study. The result is readily visible to many people worldwide, even if their only source of information is from newspapers: seldom does a month go by without some announcement of a discovery or new understanding of the universe.

One final example: the question "Can I predict the weather five days in advance?" elicits a fast answer: "No."[3] For some, that response sends them packing, with a figurative tail between their legs, to another area of research. For others, they turn the question around and ask "Under what conditions is a prediction correct five days in advance?" That alteration spurs the meteorologist to search weather records to find those situations when the five-day advance forecast turned out to be correct. He or she then investigates the conditions producing such predictability to try to understand which of several weather factors are the most critical for successful long-range forecasts. Notice that we still have not answered the original question. What is the scientific return in this example? The original question remains unanswered, yet visible progress exists. The scientific return is high.

Scientific return also depends on the context in which the question is asked. An inexpensive satellite that will survey the entire sky for the first time will be judged to have a high scientific return because a first look defines the questions to be addressed in subsequent studies. An expensive instrument merely adding the equivalent of another decimal point to an already well understood topic may provide an apparently low scientific return, assuming finite resources. Early rocket flights and satellites are examples of the former case. For the most part, examples of the latter do not exist because the proposed project does not survive peer review.

Peer review appears at least twice in a satellite's life: at the time choices are made about which projects to support and at the time choices are made in the satellite's use (after launch). Let's examine the latter first because the situation is easier to describe.

All NASA-sponsored satellites are now open to anyone who submits a proposal to use the satellite. A proposal consists of a few pages of description of what the proposer, the scientist, intends to accomplish with the requested observing time, the approach to carry out the observation and obtain the scientific result, and a short discussion of the observation's feasibility. The proposer must justify the observation by scientific return and merit.

Panels of peers, members of the community semirandomly selected,[4] review the proposals, judge them, and rank each proposal relative to the others. Conflicts of interest are avoided by sorting the proposals so that different panels review them. A knowledgeable reviewer presents a short synopsis of each proposal. The panel briefly discusses the merits of the proposal and then gives it a score. Each panel has a limited amount of time to award to the proposals in that panel. At the end of the review, the panel submits a ranked list of the proposals to NASA.

"Scientific merit" is a judgment of the quality of the proposal and its importance rendered by each panel member. As the observing time on a satellite is limited and therefore valuable, no one wants to see it wasted on frivolous ideas or investigations unlikely to yield useful science. As a result, peer review guarantees a degree of caution and mainstream thinking. That statement should not be viewed as a criticism. Anyone in a similar situation follows the same path; if the economy contracts, for example, few of us undertake spending sprees because we perceive our financial assets as more valuable and less easily replaced. We become more conservative in our spending.

For a satellite as popular as Chandra, the amount of time requested is about six to eight times the amount available. About one proposal of every six or eight will be accepted, assuming all proposals ask for the same amount of time, on average. It is a competitive process.

Just as peer review decides what science investigations will be carried out using the satellite, so, too, does peer review decide the projects to be implemented. Periodically, NASA releases a Request for Proposals for a particular flight opportunity. Interested scientists respond to the request by preparing a proposal for a particular satellite, outlining the science to be obtained, why it is important, its impact on other areas of study, its lifetime, how the satellite is to be operated, and costs. The proposals are judged, ranked, and recommended for funds by a group of scientists selected for their expertise and experience not only in the science but also in such areas as flight operations. Although the evaluation of the science is the most important factor, a team with little or no experience faces a tougher scrutiny than one with experience. The process just described applies to every satellite project approved for flight. For small and medium-sized projects, NASA management can often choose based on their budgets and the recommendations of the science-judging panel. For large projects, an additional voice often weighs in.

About once every ten years since 1964, astronomers and astrophysicists in the United States review the scientific progress in the field. These reviews occur under the auspices of the National Research Council of the National Academy of Sciences. Each review is named informally for the person who chairs the committee charged with the review.[5] A review committee, the members of which coordinate smaller reviews focused more tightly on specific questions, directs each review. The decade reviews provide essentially a "we are here" map of astrophysics. During the review period, the community develops a consensus on the big questions that need to be addressed, or questions on which substantial progress is possible, and the specific projects that will provide the necessary data. The review culminates with a list of projects for which financial support will be sought from Congress during the coming decade. The astronomical community ranks the projects by the importance of the issue;[6] that ranking is often determined to some degree by the scientific return expected from a specific project. The Chandra project ranked number one on the final list of the Field Committee in 1982; the Bahcall report in 1991 reaffirmed its rank. The added weight of a decade review can push a large project, whether satellite or ground based, onto Congress's "to do" list.

How does scientific return affect progress and the costs of hardware? Why, for example, do scientific instruments cost increasingly more money? Is the purchase of scientific instruments and detectors much like outfitting a workshop with tools? Such questions certainly interest taxpayers and legislators who appropriate the money to pay for a given program.

The word "progress" suggests an increasing understanding of the subject under investigation, be it AIDS, heart disease, cancer, black holes, lasers, or superconductors. The term "understanding" means that some questions have been asked and answered. By itself, asking and answering questions is not sufficient. Predictions must be made based on our limited understanding and those predictions must be tested against data. That cycle represents progress in science.

We seemingly have moved far from the question: why do we continually need increasingly sophisticated hardware that costs more and more money? The discussion above alludes to the answer without providing it directly: the increasingly precise theory requires more-detailed observations. More-detailed observations can be obtained only with better observational hardware: better telescopes or better detectors (or, frequently, both).

What does better hardware mean? When a researcher considers diving into a new area of research, the first question is often "Is anything there?" We saw just this approach taken in the early rocket flights. For example, an astrophysicist might ask "do any objects in the universe exist that emit X rays?" This can be an easy question to answer: launch a satellite that carries a detector that is sensitive to X rays. If the detector sees X rays, the question has an answer. If the detector collects nothing, the scientist may consider assembling a more sensitive detector and trying again. The direction, intensity, time variability, and spectrum from the source are all irrelevant. Once the scientist knows that at least one object in the universe emits X rays, then many more questions should pop into his or her mind (at least, if he or she is paying attention to the data). How many objects exist? What kinds of sources are they? How hot are they? Are they steady or do they vary? The list goes on. To answer any one of them, the scientist requires a detector that will provide more than a simple "yes/no" detection.

Let us assume we have detected X rays from a source outside our solar system. Imagine that our science goal is to focus more sharply than ever before. What is required of the hardware?

To obtain an image with higher resolution, the portions of the satellite that interact with the light from the source must be constructed to high precision. For X rays, this means polishing the mirrors to a high degree of smoothness (the "skipping stone" analogy). The surface roughness must be smaller than the wavelength of the light incident on it or the surface will deflect the incoming photons.[7] This is a quality issue: how well do you want your telescope to perform? If we want the best X-ray mirror ever placed in orbit, then we must smooth the reflective surface to a higher precision than ever before.

Are we finished? Not at all. The requirements of increased precision in pointing and the increased smoothness of the mirrors propagate throughout the entire satellite. Let's look at the mirrors again. The mirrors must be mounted in a frame that will keep them from changing their shape. If we've spent the money to create a smooth surface, it will be a wasted effort if the mirrors sag or bow because the frame is not sufficiently strong. In addition, the frame must also hold the detectors fixed in position relative to the mirror focus. These requirements translate into a higher precision in the manufacture of the pieces of the framework. Higher precision costs more.

The increased precision of the mirrors will deliver more photons to a smaller area, but we still must detect them. Recall what a telescope does: it produces in the detector an image of the object under study. To see finer details in the object, our detector must have smaller pixels to separate those details. Smaller pixels require more precision in the construction of the detector. In addition, small pixels generally mean that there are more of them, because we want to cover the largest field of view possible. Each pixel must be read, so the electronics must be capable of handling a large number of pixels. This increases the requirements on the power supply, which, for a spacecraft, means larger solar arrays.

Finally, all our efforts to meet the science goal will be useless if we do not know precisely where the satellite is aimed. To locate a source, the satellite must be equipped with gyroscopes that measure the change in position of the satellite as it moves, or high-quality star trackers to hold the satellite's direction of pointing to high precision. Either approach requires higher quality, so the cost increases.

When all these items are added together, the weight of the satellite has increased, so more fuel is required to place the satellite into orbit. The total cost of the program becomes high. Politicians and program administrators choke and urge that cuts be made.

Scientists understand that, for budgetary reasons, proposed programs are not always funded. There are always more questions and proposed instruments than money to fund them. Sometimes a program becomes so expensive that it simply cannot be funded unless it is severely "descoped" (a euphemism for "cut back").

Descoping means the elimination of some of the science goals with the aim of reducing costs.

Increasing the quality of an instrument will always increase its cost. Unfortunately and unintentionally, the peer review process had added its own cost: "mission creep." Although mission creep may sound sleazy, it is a natural response of dedicated scientists. If I, a scientist with an idea for a detector, see a proposed project that appears to be the only project an agency such as NASA intends to support and the only project apparently to receive funding from Congress, I will make an effort to attach my detector to that satellite. However, the added weight and drain on the power supply of the satellite requires a slightly larger satellite and more power. The addition of my favorite detector leads another scientist with her favorite detector to join the project. Mission creep occurs if the project becomes not only the only game in town, but also the only game for the foreseeable future.

As a result of this tendency, project costs spiral upward. During the design phase of the Hubble Space Telescope, NASA viewed the future Hubble as the only space telescope project to be undertaken; the list of proposed instruments exceeded space and weight limitations.[8] Mission creep may explain part of the complexity of Hubble. Mars Observer cost about $1 billion and was lost just before entering orbit around Mars because a transistor in its clock (the redundant crystal oscillator) failed. It, too, was viewed as the only Mars mission for the 1990s. With an annual budget of about $12 billion, NASA cannot afford to procure too many billion-dollar missions. Science suffers.

One approach toward the justification of billion-dollar missions is the Great Observatories program. In 1985, NASA program managers decided to group several missions then in the planning stages into a single package,[9] the Great Observatories. The scientific goal: fly several missions at one time to cover much of the electromagnetic spectrum. Several missions became identified as part of the Great Observatory program. The planning evolved to include four large missions: the Large Space Telescope (covering the optical band, later renamed the Hubble Space Telescope), the Advanced X-ray Astrophysics Facility (later renamed Chandra),[10] the Gamma-ray Observatory (renamed the Compton Observatory), and the Space Infrared Telescope Facility.

The value of simultaneous observations across the electromagnetic spectrum was, at that time, only beginning to be appreciated. For objects that do not vary or that vary over very long times, a scientist easily collects observations at all desired wavelengths because exact simultaneity of the observations is not required. The researcher must learn as much as possible about the instruments used in the various bands or collaborate with others who work in the other bands. The only requirement for the researcher is the patience to collect all the data.

For variable objects, however, simultaneous observations are absolutely necessary. If a burst of light occurs in the ultraviolet part of the spectrum, we want to know whether that burst also occurred in the X-ray, optical, infrared, and radio bands. Only simultaneous observations provide the answer.

Why multiple observatories? First, we must launch different satellites because no one method to detect photons works in all bands. Consider just the mirrors; we've learned that in the X-ray band, grazing-incidence mirrors are required; in the ultraviolet, optical, and infrared bands, normal-incidence mirrors are standard. Second, we need the satellites to be launched at about the same time and to live for about the same length of time. Third, we want to be able to schedule observations with each satellite individually or with all of them simultaneously to provide maximum flexibility. Each satellite is then relatively simple to use (little mission creep), with the complexity reserved for the coordination of the satellite constellation.

Unfortunately, the Great Observatory program ran headlong into the budget limitations of the early and mid-1980s, thereby limiting the multisatellite experiment, at least for the foreseeable future. Each Great Observatory, as designed, arrived with a billion-dollar price tag, pushing other proposed satellites off the budgetary table. Attempts to launch all four observatories nearly simultaneously simply could not be accomplished within the budgets allocated. Science suffered.

Dan Goldin, appointed the administrator for NASA by President George H. W. Bush in February 1992 , fought this trend with his mantra "faster, better, cheaper."[11] If a project is not stretched out over many years, then the project is immediately less expensive because the amount spent on salaries is lower, salaries of project workers being among the largest items in a project's budget.[12] Scientists are happier with the "faster" part because they get to see the scientific results of their concept.[13] "Faster" also means that the progress in understanding the universe occurs at a higher rate. "Cheaper" and "faster" imply that more missions can be attempted.[14] The original design for the Space Infrared Telescope Facility (SIRTF) cost about $2.2 billion; the project has since been redesigned (descoped) yet it retains about 80 percent of its original capabilities, largely because the complexity of the instruments was reduced and the nature of the orbit was changed.[15] The program now costs about $450 million. The ten years from about 1997 to 2007 have seen and will see some six to eight missions to Mars.[16] Compare that with the twenty years from 1975 to 1995, when there were exactly three missions to Mars: Vikings 1 and 2 in 1976, and Mars Observer in 1993.[17] Twenty years is a long time to wait between asking a question and obtaining an answer.[18]

Along with the "faster, better, cheaper" mantra came a scientific coherence organized around grand questions and tied to the decade reviews. The Office of Space Science at NASA has a well-defined set of goals focused on four themes: origins, structure and evolution of the universe, exploration of the solar system, and the Sun-Earth connection. Within each theme, NASA managers and scientific advisors created a set of fundamental goals. For each goal, they also enunciated a few specific investigations, each leading to a specific space mission. As a result, satellite projects will generally have more sharply focused goals than in previous years.

What are these themes? The exploration of the solar system theme is clear from its title. For the Sun-Earth connection, the goals focus on tracing the flow of energy from the Sun and determining its effects on the solar system, Earth, and humanity. The effects include climate change, impacts on power and communication systems, and a comparison of Earth's environment with those of the other planets.[19] The last two themes are the structure and evolution of the universe and origins. The goals are the determination of the size and geometry of the universe, the tracing of the evolution of the chemical elements throughout the complete stellar cycle, and the exploration of the extreme physics of gravity.

Along with each theme comes a set of questions. Any mission proponent striving for support from NASA, and, ultimately, from the United States Congress, must demonstrate how its proposed mission intends to address one or more of the fundamental questions. In this way, the proposers must consider the proposed mission's importance relative to other missions and its potential impact on one of the themes. Essentially, the reorganization has produced a road map of NASA's aims for the next decade or two.[20]

Earlier, we discussed why the cost of telescopes (as well as accelerators and other scientific instruments) continually increases. We considered only one hardware property, that of spatial resolution. What about the others: time resolution, spectral resolution, and sensitivity?

Increasing the spectral resolution may be the most difficult item on the list. For all of the others, the increase can often be obtained by doing more of the same: adding more grazing-incidence-mirror shells, for example, will increase the sensitivity by increasing the collecting area. Increasing the spectral resolution often requires completely new detector science. The limits of proportional counters forced astrophysicists to examine CCDs as potential X-ray detectors. The change from proportional counters to CCDs was a change in the physics of how we detect an X ray. We have already reached the limits of CCDs. This limitation has pushed researchers in the direction of X-ray calorimeters.

Increasing the time resolution requires faster electronics, so the satellite must be capable of handling a considerably larger data load. If collected faster, the data must be handled rapidly and quickly sent to the ground or stored on board the spacecraft. If the data are stored onboard, then the data recorders (in the old days, actual tape recorders;[21] now, solid-state memory is used) must have a larger capacity so they do not overflow, causing the loss of valuable data. If transmitted to the ground, the data must be packed into a stream and sent to another satellite or to a ground station.[22]

To date, one satellite has been dedicated to high-time resolution studies of the X-ray sky: the Rossi X-ray Timing Explorer. This satellite has computer-processing units dedicated to collecting and to processing the data on board with a time resolution of a millionth of a second.[23] RXTE also can make use of NASA's network of data-relay satellites, which are capable of megabit-per-second data rates.

These rates are rather high for astronomy satellites (Chandra's data rate is 25 kilobits per second) because RXTE carries the largest proportional counters ever flown. Consequently, bright sources lead to huge data rates with RXTE: the brightest source collects about a quarter of a million X-rays per second.

To increase the sensitivity requires a telescope with one of the following improvements: a larger collecting area, precision optics producing a smaller point-spread function, or detectors with a much lower internal background. The ideal telescope would include all three properties. As described earlier, larger mirrors means the entire spacecraft increases in weight. Detectors contribute noise to the measurement process; if hardware builders reduce that noise contribution, then every detected event may be a real X ray and not a false signal introduced by the detector itself.

Increasing the collecting area by a factor of one hundred means that nearby X-ray sources may be investigated in great detail and faint, distant sources studied for the first time. Any observational science requires as many examples of behavior as possible. More examples also often cover a broader range of behavior. With a broader range of behavior, we test our understanding under different conditions. Unique sources do not usually allow us to increase our understanding because we often do not know whether the unique behavior is critical or just a trick of nature.

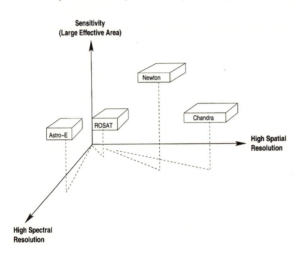

We now finally address the opening question, "Why three satellites?" The designers of each satellite focused on slightly different instrument parameters. Figure 9.1 shows the situation for three (spatial, spectral, sensitivity) of the four parameters (time, the fourth, is left out). Chandra has a collecting area about four times as large as the Einstein Observatory, but a point-

Fig. 9.1. The spatial-spectral-sensitivity niches occupied by the satellites Astro-E, Newton, and Chandra, as well as ROSAT; for Chandra and Newton, the relative positions on the spectral and sensitivity axes represent the spectral resolution and effective area of the primary instruments without the gratings. If the gratings were included, the two boxes representing Chandra and Newton would slide outward on the spectral-resolution axis, but downward on the sensitivity axis, because the effective area with the gratings inserted is much lower. The positions are relative and serve to illustrate the complementary nature of the three satellites. The impact of the loss of Astro-E is clear: the high spectral-resolution axis is essentially devoid of coverage. The ideal X-ray observatory exists, within this three-axis space, at about the location of the reader's eyes: very high spectral resolution, very high spatial resolution, and very large effective area. A fourth axis, high time resolution, has been suppressed to avoid attempting to flatten a four-dimensional figure to two-dimensional paper.

spread function about a factor of 10 better than ROSAT and about a factor of 120 better than Einstein. This combination puts Chandra high on the spatial-resolution axis. Newton, the European X-ray counterpart, has an effective area about twice as large as Chandra, but a point-spread function about 10 to 15 times as wide.[24] For bright objects, Newton obtains higher signal-to-noise spectra in a shorter time than Chandra. A larger point-spread function also includes more X rays from the background, so the minimum number of X rays to detect a faint source must be higher for Newton than for Chandra. At the faint end, then, Newton's larger collecting area is somewhat offset by Chandra's sharper point-spread function. The designers of Astro-E pursued high spectral resolution. The three satellites were to complement each other nicely because they represented opposing trade-offs for about the same amount of money.

Scientists are continually asked to justify the costs of their research, not only to funding agencies but also to Congress and the taxpayer. Astrophysics is arguably the last of the pure sciences for which the commercial opportunities hover quite near zero. How do we justify spending money on esoteric questions in the realm of astrophysics?

Unfortunately, the person considering the bottom line searches for salable products or innovations. NASA publishes lists of spin-offs, technical products that flow from research or development occurring at one of its centers. Spin-offs are just that, products that fall away from the research and development efforts for aeronautical or astronautical endeavors. Spin-offs: serendipitous results with little prior expectation.

The fundamental justification for investment in space science, astronomy, and astrophysics is education. Few other sciences attract the interest of people of all ages as much as astronomy and astrophysics do.[25] A technological society such as ours requires scientifically and technically literate citizens. Presentations of astronomical and astrophysical results by professional astronomers form one avenue by which citizens connect with science and technology. Students, future citizens, respond enthusiastically by participating in forefront research, broadening their understanding of the questions scientists ask and answer. A fraction attend college and graduate school in a technical field to form the next generation of scientific explorers.

10

"I like the dreams of the future better than the history of the past."

—Thomas Jefferson

How did the universe start? How did life start? How will it end? When will it end? We've asked these questions since childhood; the questions remain with many of us throughout our lives. Fifty years ago, none of them could be answered; 50 years from now, at least one of them will have an answer.[1] Fifty years ago, we had not yet opened the X-ray window to look at anything other than the Sun. Our knowledge of the astronomical sources of gamma rays was virtually nonexistent. Radio astronomy, recovering from the years of relative forced neglect caused by world war, opened the window on the basic structure of our galaxy as well as uncovering extragalactic objects, some of which showed no obvious optical counterparts.[2] The ultraviolet window opened only with the advent of routine access to rockets and satellites. Our quest to understand the universe now involves essentially all wavelengths.

Does the study of the X-ray universe end with Chandra or Newton? The answer is a definite "no." American scientists already plan a successor to Chandra: Constellation-X, or Con-X, originally called HTXS, for High Throughput X-ray Satellite. The phrase "high throughput" is an engineering term describing the fraction of the input signal that actually reaches the scientist. The basic design goal calls for an increase in the effective collecting area of the mirrors by a factor of at least one hundred. For X-ray astrophysicists, high throughput leads to the detection of more X-ray photons at all energies. As we have seen, the available collecting area limits our ability to collect the data necessary to push back the frontiers of ignorance.

Ground-based astronomers routinely access a set of large optical telescopes unavailable just a few years ago: the two 10-meter Keck telescopes (Hawaii), the 6.5-meter MMT (Arizona),[3] the four 8-meter telescopes for European astronomers (Chile), the two 8-meter Gemini telescopes (Hawaii and Chile), as well as a few other large telescopes still under construction. Whereas Chandra is the equivalent of the Hubble Space Telescope, Constellation-X will be the X-ray equivalent of one of the new 8-meter-class optical telescopes.

The design of Constellation-X steps off the path taken by all previous imaging X-ray satellites. The effective area of Con-X will not reside on one megasatellite but will be distributed across a constellation of several satellites. The original plan called for six or seven satellites to be launched over a series of months. Technical assessments in the past few years have demonstrated that two,[4] three,[5] or as many as six separate satellites (Fig. 10.1)[6] will provide the required effective area. The observatory will operate by coordinating the separate satellites in tandem, hence the "constellation" portion of its name. Multiple satellites will be only marginally more expensive to construct than one satellite because assembly-line concepts become applicable. Communications satellites, such as the less-than-successful Iridium satellites, have been built with assembly-line efficiencies, but science satellites have always been one-time projects. With an assembly line, the lessons learned on satellite number one improve satellites later in the series. Material purchased in bulk lowers per-item costs. With systems-engineering principles employed, the marginal cost of one more satellite becomes small. A new

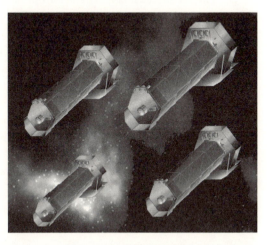

Fig. 10.1. Artist's sketch of Constellation-X, an observatory in the planning stages for launch late in the first decade of the twenty-first century. The planned observatory is composed of several independent satellites that are to be operated as a unit. (Image obtained from the public archives at the High Energy Astrophysics Science Archive Research Center, NASA-Goddard Space Flight Center.)

concept emerges: grow the effective area of the observatory. The observatory becomes operational as soon as the first satellite reaches its orbital position. Operational procedures are debugged with one satellite before subsequent satellites leave the ground.

For those who devote several years of their lives to placing a satellite into orbit, there is an additional benefit: the loss of one satellite to a launch failure does not cripple the observatory. When your job depends on a rocket launch, you certainly focus your attention on the launch.

The Con-X design team faces two challenges. First, each satellite must be identical to the others so that the data can be combined. Several recent X-ray satellites have featured multiple telescopes housed in one satellite. ASCA is an example: the CCD spectrometer actually consisted of two spectrometers, each fed by a separate set of mirrors. Spectra from spectrometer one could not be added to the spectra from spectrometer two unless the signal-to-noise ratio was sufficiently poor that the differences between the two telescopes contributed few errors in the resulting sum. To analyze the data, scientists simultaneously fit a model to the data from each detector separately. That approach is not so effective as combining data because the poor signal-to-noise ratio of each individual spectrum masks the accurate definition of emission lines, for example. Combining data properly drives down the noise, consequently increasing the signal-to-noise ratio. Emission lines become more sharply defined, particularly where the line joins the continuum.

The second challenge is that of operating the entire group of satellites. You might think that coordinating a group of satellites is a problem solved by the space communications industry. That industry, however, generally does not move the entire group of satellites at one time. In fact, most communication satellites are parked in geosynchronous orbit and remain in one location relative to a spot on Earth.[7] To aim the Con-X observatory at a different target, the entire group must be reoriented.

Con-X focuses on spectroscopy. Following Chandra, and assuming it lasts well beyond its design lifetime, we will know which sources are pointlike and which are extended for a large fraction of the brightest approximately 20 thousand sources. We will also have good images of the brighter extended sources. Newton will provide spectra for sources that are too faint for Chandra, raising the overall census of spectra of X-ray emitting objects. We will not have high-resolution spectra for a large fraction of these sources. Chandra and Newton contribute grating spectra for the brightest sources of each type, but the list of available targets for grating spectroscopy will be exhausted relatively quickly simply because the effective areas of both satellites, while large relative to previous missions, still remains small for observations using a grating.[8] Progress in science most often comes from the accumulation of evidence, not the oddball object. To accumulate evidence of sufficient weight requires numbers of objects.

Consider the following example. There are approximately 25 known "intermediate polars," a type of magnetic cataclysmic variable. Novae and dwarf novae are two types of cataclysmic variable. Intermediate polars are a third type, for which the magnetic field of the white dwarf is sufficiently strong to control the flow of matter accreting onto the white dwarf, but too weak to disrupt the accretion disk. The detailed X-ray emission from intermediate polars shows a broad range of properties. Some of the binaries suffer strong absorption at lower energies, while others show relatively little absorption. Several binaries show evidence for strong emission lines at 6.6 keV from iron, while others have at best weak emission lines. Some of this range of behavior is presumably related to the strength of the magnetic field: whether sufficiently strong to disrupt much of a disk, or weaker, leaving most of the accretion disk intact. Unfortunately, if one believes that there are subclasses of behavior exhibited by the collection of available spectra (mostly from ASCA), there are too few binaries to establish the case with any reliability. With so few systems to study, detecting overall trends becomes difficult. Science suffers.

With the current observatories, we can study about a third of all the intermediate polars with sufficient spectral detail to see broad trends in the spectroscopy. We can barely study the time-dependent behavior because the signal-to-noise ratio is too low to allow us to subdivide the data by time. Con-X would eliminate these limitations.

The large effective area of Con-X opens several new dimensions of study. Time-resolved spectroscopy is one. Another is the evolution of the chemical elements with redshift. Con-X also opens the door to hard-X-ray imaging. Essentially no satellite has generated an image of the sky with good spatial resolution better than about 1-arc minute above 10 keV but below 50 keV.[9] For the study of active galaxies, spectroscopy at energies above 10 keV is necessary to ascertain the legitimacy of the unified theory of active galaxies.

There are many types of active galaxies: radio-quiet and radio-loud quasars, broad-absorption-line quasars, types 1 and 2 Seyfert galaxies, Narrow Line Seyfert galaxies, blazars, optically violent variables, and ultraluminous infrared galaxies.[10] To the nonspecialist, the list of behaviors is confusing and constitutes a veritable zoo. These behaviors, however, may turn out to be the key to our view of the nuclear region, the viewpoint or unified model of active galaxies we encountered previously. The model describes an active galaxy as a galactic black hole surrounded by a torus of matter from which it accretes. The accretion torus generates jets that eject matter and radiation parallel to the axis of the torus. An edge-on view cuts directly through the thickest parts of the torus; the absorption is so high that X rays up to 5 and 6 keV are blocked. The model predicts that some active galaxies remain to be discovered because these galaxies are tipped so that we view them essentially edge-on through the torus.

If we sum the contributors to the total energy density (energy per unit volume) of the X-ray background, the peak occurs at about 30 keV. This suggests a separate type of object, emitting above 10 to 20 keV or so, but little below those energies. The viewpoint model for active galaxies naturally explains this new class of object: these are the active galaxies viewed edge-on. When estimates are calculated, astronomers realize that potentially large numbers of active galaxies may exist that are not currently detected.

Only a telescope with a broad bandpass can study these obscured galaxies. RXTE has the bandpass, but it does not have the spectral resolution. Observations with RXTE and Chandra can be coordinated to observe the same object simultaneously, but the coordination of two satellites in different orbits is difficult. Con-X will provide not only the bandpass, but also the spectral resolution in the 10–40-keV band to locate these active galaxies, all in one observatory.

The European community recently chose a successor to Newton: XEUS (Fig. 10.2 [see next page]).[11] XEUS is an ambitious project: to place about six square meters of effective imaging area into orbit with an imaging resolution of 2 arc-seconds at

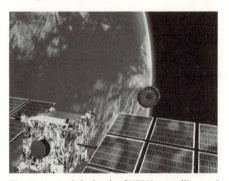

Fig. 10.2.   Artist's sketch of XEUS, a satellite on the drawing board. Note that XEUS comes in two pieces, the mirror portion (the most distant piece) and the detector portion. XEUS would be partially assembled at the International Space Station. (Image used by permission of the European Space Agency's XEUS Study Scientist, Dr. M. Bavdaz.)

1 keV and spectral resolution of between 1 and 10 eV across the entire XEUS band of 0.05 to 30 keV. Six square meters equals 60,000 square centimeters; Chandra has an effective area of about 600 square centimeters; Newton an effective area of about 1,200 to 1,400 square centimeters. Scientists designed Con-X to reach 10,000 to 15,000 square centimeters. The ambitiousness of the project stems from the many substantial factors of improvement that must occur before construction commences. Current spectral resolution for calorimeters is closer to 10 eV than to 1 eV; improvement by a factor of five is likely in the coming five or so years. That is the easy part.

To place six square meters of mirror into orbit requires new approaches to mirror construction. Chandra's mirrors are glass, maintained in position by a substantial support structure. The combination of the thick glass and the support structure prevents the close stacking of additional shells. Close-packed shells increase the effective area of the telescope. One solution, flown on the Astro-1/BBXRT shuttle mission and on ASCA, used thin foil mirrors.[12] Unfortunately, the spatial resolution suffered because of inherent defects in the quality of fabrication as well as in the mounting of the foils. Attempts to improve the fabrication quality have not enjoyed substantial success. The foil is simply not sufficiently smooth, nor can the many foils be aligned sufficiently well, to produce a sharp point-spread function. Even the use of glass for the mirrors, such as on Chandra, is not an option because of its substantial weight. Finally, the telescope must be launched. The size of the rocket used to launch the satellite limits the size of the mirror of the telescope, unless clever alternatives are designed. Six square meters requires a mirror at least four to five meters in diameter, about the diameter of the payload housing of the largest rockets. An early design builds the mirrors using "petals," similar to those in a flower bud. In orbit, the petals bloom and mimic the required shape of grazing incidence optics.

An increase in effective imaging area also propagates through the entire design of the telescope. As the diameter of the mirrors increases, the focal length must increase. The focal length of X-ray telescopes, the distance from mirror to focus, is long because of the grazing incidence.[13] The size of payload bays on rockets or shuttles requires limitations not only on the diameter of telescopes, but also on their focal length. Chandra is about as large as it could possibly be and still fit within the space shuttle's payload bay. The flying version of ASCA was longer than the payload volume of the Japanese rocket used to launch it; to fit the satellite into the rocket, Japanese engineers designed the mirror support and infrastructure to extend, or unfold, after reaching orbit. For XEUS, the focal length is about 50 meters (more than half a football field). Instead of attempting to assemble a spacecraft of that length, a XEUS study team has proposed two spacecraft, with the mirror portion flying 50 meters in front of the detector portion. To maintain a sharp focus, the two spacecraft must remain fixed relative to each other to within about one millimeter. "Formation flying" has become all the rage, driven by scientific necessity.[14]

The ambitious designers of XEUS were not content with just six square meters of observatory. The "petal" plan includes the possibility of adding rings of petals after the spacecraft reaches orbit. The six square meters of mirrors will be used for several years as the mirror team prepares additional petal rings. Once the hardware is ready, flight controllers will maneuver XEUS to the International Space Station, where additional rings will be added, totaling as much as 30 square meters. The space station offers some intriguing advantages. Final steps of assembly could be undertaken in space, for example, and refurbishment or maintenance possibilities exist. With careful planning, the observatory might last decades rather than years.

The goals for XEUS are very similar to those for Constellation-X: increase the number of detected X rays across a broad range of energies to peer deeper into the X-ray-emitting regions of the universe. The increased effective area will permit the probing of individual X-ray-emitting objects in nearby galaxies. The gas and dust in the equatorial plane of our galaxy masks many soft-X-ray sources. The study of external galaxies provides an outsider's view of galaxies we judge to be similar to our own, a view that is simply not available to us because of the high absorption within the plane of our galaxy. For clusters of galaxies, studies of the hot gas will be possible to high redshift, thereby letting us probe the evolution of the largest cosmic structures back to the days when those structures were young.

The conceptual design for XEUS is ambitious, and, as with all ambitious targets, the lure of potential achievement motivates the effort. If the Europeans do succeed, they will have an X-ray telescope capable of carrying out the types of research only dreamed about by today's researchers.

Not to be outdone, several researchers at NASA have proposed a mission with 150 square meters of collecting area with an image quality of one second of arc (just slightly worse than Chandra's image quality). The stated goal is a detection sensitivity limit ten thousand times better than Chandra's, or about 10^{-20} erg s^{-1} cm^{-2}. The satellite's working name is Generation-X, and it is proposed for about 2020.

What might we learn from Chandra, Newton, Con-X, and XEUS over the coming years?[15] Speculating on the answer to such a question is anathema to any scientist familiar with the topic precisely because we cannot include surprises. No one predicted X-rays from comets or elliptical galaxies or protostars. A future path that seems relatively predictable can change nearly overnight with new observations or new understanding.

Studies of the X-ray emission of normal spiral galaxies may be one of the fields of astrophysics research benefiting most from observations from Chandra. X-ray-emitting objects such as stars, supernova remnants, and X-ray binaries combine to form galaxies. Spiral galaxies also contain an interstellar medium; essentially, the space between the stars is filled with various quantities of gas and dust. The interstellar medium has several phases: a cold phase, best studied in the radio band; a cool phase accessible to infrared telescopes; a warm phase, which can be studied in the optical and ultraviolet bands; and a hot component, best detected in the far-ultraviolet and X-ray bands. As a galaxy forms, its size, composition, and separation from its neighbors influence its evolution. The flow of matter within a galaxy, from gas clouds to stars to supernovae to remnants and back to gas clouds, is confined to the galaxy to a fair approximation, assuming that the galaxy is not colliding with a neighbor.

The matter flow loops from cold, dark clouds of gas and dust in the interstellar medium to stars converting hydrogen to chemical elements of higher atomic mass (such as carbon, nitrogen, oxygen, neon, silicon, and iron) by nucleosynthesis. Much of the synthesized matter is returned to the interstellar medium during a star's last years, when it loses considerable mass or when it explodes. Because stars manufacture heavier chemical elements from lighter ones, the overall "metallicity" of a galaxy increases. Metallicity is an astrophysicist's term for the fraction of the total mass of a star or galaxy that is composed of "metals." (Metals, to astrophysicists, are all chemical elements except hydrogen or helium.) Essentially, then, the metallicity measures the amount of "pollution" that stars have produced. Before anyone becomes agitated by the idea that stars have polluted the universe, we must remember that without that pollution, we would not exist. A universe without polluting stars would be a universe of hydrogen soup with a helium flavoring tossed in. No carbon. No nitrogen. No oxygen. Without CHON (carbon, hydrogen, oxygen, nitrogen), no plants, no life.

For galaxies, then, we have relatively simple goals. We want to compare the objects we observe in our galaxy with similar objects observed in other galaxies. The launch of the Einstein Observatory in 1978 gave us our first look and told us that spiral galaxies emit X rays and that some of the X rays probably came from specific objects within a galaxy. That conclusion arose from observations of galaxies that uncovered luminous X-ray-emitting objects located off center, or multiple luminous sources. The spatial resolution was not sufficiently good to study, in detail, more than a handful of the nearest spiral galaxies. ROSAT provided better spatial resolution, and with it, astrophysicists studied about two dozen spirals. About two hundred spiral galaxies exist within 30 million light-years, but we cannot describe the appearance in the X-ray band of most of these neighbors. This is a somewhat embarrassing, if not shameful, situation.

With the spatial resolution of Chandra, however, observations of these galaxies will be straightforward. Furthermore, Chandra's spatial resolution essentially matches the resolution typically obtained at other wavebands, particularly the optical and radio bands. We will compare the data across wavebands to assess, for example, which galaxies possess hot components of their interstellar media and which do not. Is there a correspondence or correlation between galaxies with hot components and, for example, the total mass of a galaxy? Is the activity of X-ray binaries in our galaxy similar to the activity of X-ray binaries in other spirals? How do the X-ray-emitting populations of the galaxies in various environments (galaxies in clusters versus isolated galaxies) compare? We do not know the answers to these questions today. By the end of the Chandra and Newton missions, we will have a solid survey of our two hundred neighboring galaxies and will at least attempt to answer some of those questions.

Answering questions about galaxies will tell us more about the abundances of chemical elements. The hot phase of the interstellar medium is relatively unexplored because it requires telescopes with large effective areas plus at least good spatial resolution. Chandra and Newton complement each other well: Chandra's sharp point-spread function locates the point sources, while Newton's larger effective area collects the photons from the faint, diffuse gas. It may be possible in a few years to use the Chandra observation of a galaxy to deconvolve, or remove, the point-source population from an observation of the same galaxy with Newton and analyze the two types of X-ray emission separately. The amounts of metals increase with time because of the processing of the elements within stars. We learn about the rate at which the amounts of the elements change by looking at galaxies that are farther and farther away.

Remember that the farther away an object is in the universe, the further back in time we look. "Look-back time" is actually easy to understand. Imagine communicating with a friend living on the opposite side of the country only by posted letter ("snail mail"). Your friend writes a letter on Sunday and drops it into a mailbox. You read it on Friday, but you are reading about events that occurred five days earlier. An astrophysicist would say that you have a "look-back time" of five days. In other words, the letter required five days of travel to reach you. The universe behaves the same way with information traveling at the speed of light.

Applying this idea to the measurement of chemical elements, we want to obtain spectra of objects as early in the history of the universe as possible. We expect a spectrum from such an object to show fewer emission lines of metals. If it does not, we must revise our ideas of how the universe evolves. X rays are particularly important because most of the visible mass of the universe resides in hot gas. The farther out we look, the more blended are the individual sources (because we cannot make an infinitely sharp point-spread function). We switch from studying individual objects to studying an ensemble of objects as the only sensible alternative. Clusters of galaxies are uniquely suited to the study of the evolution of the elements: as we've learned, clusters are hot spheres of gas. That gas represents an average of the individual components over the life of the cluster. X rays provide the only wavelength band that can accurately assess the amount of hot gas in clusters as we look from nearby clusters to clusters very distant from us.

On the long-range horizon lies another potential mission that truly grabs the imagination: MAXIM, the Micro-Arc-Second X-ray Imaging Mission. Earlier, an arcminute was defined as one-fifth of the separation between two of the columns of the Lincoln Memorial as depicted on the reverse of a United States penny. One arc-minute contains 60 arc-seconds. A micro-arc-second is one millionth of an arc-second. This is an extremely small angle. If we had eyes capable of reading at the micro-arc-second level, we could see a dime on the surface of the Moon. We'd also have little need for microscopes to study the structure of a leaf, for example, because we'd see many of the details of a leaf right down to its pores.[16]

How can we resolve the X-ray emission from objects at the micro-arc-second level? MAXIM will carry out its imaging using the principle of interferometry.[17] Interferometry combines the light from several smaller telescopes to simulate the resolution of a single mammoth one; this technique is known as aperture synthesis and is widely used by radio astronomers daily. The Very Large Array (VLA) in New Mexico, part of the National Radio Astronomy Observatory, is one such instrument. Built in the shape of an inverted Y, the VLA, in its largest configuration, acts like a telescope with a diameter of 35 kilometers (about 22 miles). There is no magic here: the array of telescopes does not provide the collecting area of a telescope 22 miles in diameter, only its spatial resolution.

All telescopes, regardless of the band for which they are designed, suffer a limitation based on their size and the wavelength regime. This limit is known as the diffraction limit; it sets a fundamental bound on the highest possible spatial resolution achievable with a specific telescope operating in a particular bandpass. The limitation depends on the wavelength of light divided by the size of the telescope.[18] The Hubble Space Telescope operates close to its diffraction limit: it works at wavelengths from 1,200 to 10,000 angstroms and is 2.4 meters in diameter. The diffraction limit becomes about 0.1 seconds of arc at the long wavelength end; at the short end, the diffraction limit is about a factor of 10 below Hubble's capability.

Note the definition of diffraction limit again: wavelength divided by the telescope diameter. Extremely small angular resolutions are more easily achieved in the X-ray band than at other wavelengths because the wavelength of the light is so small. As the wavelength decreases, the size of the aperture required to reach a specified angular resolution decreases. The VLA reaches an angular resolution of about 0.2 seconds of arc at a wavelength of 2 centimeters. Chandra has an angular resolution of about 0.5 seconds of arc with a diameter of about 1.2 meters; Chandra does not operate even near the diffraction limit for 1-keV X rays.

Fig. 10.3. Artist's sketch of MAXIM, a satellite on the drawing board for launch around 2020. This satellite is appealing to anyone who learns its goal; that of obtaining an image of a black hole. It makes use of X-ray interferometry. (Image used by permission of Dr. Webster Cash.)

Designs for MAXIM use the concept of formation flying similar to the XEUS design. A set of spacecraft, acting as the telescope, lies about five hundred kilometers in front of the detector spacecraft (Fig. 10.3). Lasers maintain the relative separations to about twenty nanometers (10^{-9} meter) or better. Such a spacecraft represents a substantial challenge in pointing and control; to explore the complexity, astronomers proposed a pathfinder mission, one with a goal of milli-arc-seconds (10^{-3}) rather than micro-arc-seconds (10^{-6}). The pathfinder mission could not image a black hole, but it could certainly image nearby stars. The possibility of such data intrigues stellar astronomers because they could study the behavior of stars other than the Sun. The pathfinder mission may launch near the end of the decade.

So what could we do with a mission like MAXIM? MAXIM could be used to study specific X-ray-emitting objects in nearby galaxies, or to see a supernova explode and watch its shock wave engulf a cloud. But the scientific bait, the prime scientific goal to attract interest and funding, is the ability to obtain an image of the region around a black hole. There are few concepts in astrophysics that attract more questions than the idea of black holes. The study of the X-ray universe started with a rocket flight in 1962. As I write these words, the fortieth anniversary of that flight approaches. Fewer years exist between where we are now and a MAXIM image of a black hole than lie between that first rocket flight and us. In the historical equivalent of a flash of light, we will have stepped from knowing nothing about the X-ray emission of the universe to holding a picture of a black hole. Enjoy the ride.

Notes

Preface

1. I am not alone in making this assessment. Martin Turner, a professor of astrophysics at the University of Leicester, and his coauthors presented a similar sentiment in 1997 at the Next Generation of X-ray Observatories workshop. In an article describing XEUS, the European next-generation X-ray observatory (presented in chapter 10), he wrote, "Currently approved missions like XMM, AXAF, Astro-E, SpectrumX, and ABRIXAS will complete what may be regarded as the discovery phase of X-ray astronomy." (In this book, we will meet XMM [Newton], AXAF [Chandra], and Astro-E, but not Spectrum X and ABRIXAS, which are primarily Russian and German missions, respectively.) See M. Turner et al., *Proceedings of the Next Generation of X-ray Observatories Workshop,* ed. M. Turner and M. Watson (published online, 1997).

2. For example, astrophysicists only recently discovered that comets emit X rays. No one had predicted that.

Overview

1. Yes, radio light. You cannot hear radio light any more than you can hear optical light.

2. More about waves, photons, energy, and the electromagnetic spectrum appears in chapter 2.

3. The Chandra X-ray Observatory was named for the Indian American astrophysicist Subrahmanyan Chandrasekhar (1910–95). "Chandra" was Chandrasekhar's nickname, and it means "luminous" in Sanskrit. Chandra, before NASA renamed it in December 1998, was originally known as the Advanced X-ray Astrophysics Facility (AXAF). Why "Facility" instead of "Observatory" or "Telescope"? I do not know, but sometimes I think people arrange words to achieve an easy, and memorable, acronym. "AXAT" sounds like the name of an Egyptian god. "AXAO" sounds like the noise project managers make when their budgets are cut.

4. The Hubble Space Telescope was named for the American astronomer Edwin Hubble (1889–1953).

5. The Compton Gamma-ray Observatory was named for the American physicist Arthur H. Compton (1892-1962). Compton was deorbited in June 2000 over concerns that it might come down in an uncontrolled manner.

Chapter 1

1. As this book goes to press, Japan and the United States have agreed to re-fly Astro-E using spare parts from the original. The launch of Astro-E2 is currently scheduled for early 2005.

2. The answer to this question is easy: no.

3. This date I thought could not be more auspicious, falling as it did on the thirtieth anniversary of the Apollo 11 landing at the Sea of Tranquillity.

4. The countdown essentially cannot be stopped once the solid rocket boosters are ignited. At T–7 seconds, there were only a few seconds left before the solids ignited.

5. The ignitors are the "sparklers" visible just before the main engine erupts in flame.

6. The likely gas leak suggests that the replaced sensor was correctly detecting increased vapor.

7. The point-spread function will be described in chapter 3.

8. "Cas" is short for Cassiopeia; the "A" designates the object as the first radio source observed in the constellation Cassiopeia.

9. The primary historical records are Chinese, followed by Korean and Japanese. Chinese lunar-eclipse records date to about 1200 B.C.; solar-eclipse records to about 720 B.C.; astronomical records, including records of supernovae, novae, comets, meteors, and aurorae, to about 200 B.C. The Korean and Japanese records are less extensive earlier than about A.D. 800, but are detailed after that date. One of the mysteries of the historical record is the apparent absence of European or Middle Eastern records. Additional information is available in, for example, F. Stephenson, *Applied Historical Astronomy,* Joint Discussion 6, Twenty-fourth Meeting of the International Astronomical Union, Manchester, England, August 2000, and D. H. Clark and F. R. Stephenson, *Historical Supernovae* (New York: Pergamon Press, 1977).

10. For the advanced reader: the discussion in the next few sentences is a summary of the results reported in *Astronomy and Astrophysics* 365 (2001): L202, by K. Dennerl et al.

11. N157B is an entry in the "nebula" portion of a catalog published by K. Henize in *Astrophysical Journal Supplement* 2 (1956): 315.

12. The objects have hard, power-law X-ray spectra, typical of the nuclei of active galaxies. For now, think of a hard spectrum as possessing a higher temperature than a soft spectrum. In addition, active galaxies show strong absorption at low energies. If active galaxies are essentially uniformly distributed across the sky, then we expect a few active galaxies to appear in every image of the X-ray sky, forming a population of background sources. Those sources may or may not be detected, depending on the nature of the emission of the foreground object. Galaxies such as the Large Magellanic Cloud are relatively transparent to X rays.

13. For advanced readers: this short discussion is based on a paper by M. Turner et al., *Astronomy and Astrophysics* 365 (2001): L110.

14. For those who prefer black humor and do not think about the thousands of people who spent long hours assembling the satellite, the launch inserted Astro-E into a "geostationary submarine orbit."

Chapter 2

1. Actually, there are five, but for now, we'll ignore the fifth: polarization.

2. ROSAT is short for Roentgensatellit and honors W. Roentgen. It was a collaboration among Germany, the United Kingdom, and the United States. Launched in June 1990, it lasted until 1997.

3. Roentgen is pronounced "Rent-gen."

4. The material on W. Roentgen's experiments comes from an excellent summary in *Physics Today* 48, no. 11 (November 1995), H. Seliger, on the hundredth anniversary of the discovery of X rays.

5. The words in quotes were in quotes in H. Seliger's article. I assume they came from von Helmholtz's original paper, which was published in German.

6. The "late for dinner" phrase is similar to the opening sentence in H. Seliger's article. It's too good not to use it here.

7. This comparison will become increasingly bewildering to future generations as televisions are constructed of flat panels using liquid crystals or plasma displays rather than

picture tubes. Television picture tubes are essentially small electron accelerators. A black-and-white television contains one electron beam; a color television has three beams, one for each primary color. Electrons are emitted by the cathode, which is physically hot, into the vacuum of the tube. The cathode is negatively charged, so electrons leaving its surface are attracted toward the positively charged anode. They first must pass through a small hole in a negatively charged grid, which shapes the ejected electrons into a narrow beam. The beam is deflected up-down and left-right by magnets, after which it strikes the chemical phosphor that coats the back of the television tube's curved glass face, forming the image. (Based on a discussion in Louis Bloomfield, *How Things Work* [New York: John Wiley and Sons, 1997].)

8. High voltages produce X rays; modern television tubes use a lower voltage, so X-ray production is negligible.

9. Every scientist facing the question of "What applications does your research have?" dreams of such a case of new physics and visible, immediate applications. If only all scientific discoveries could have such clear payoffs, writing grant applications and urging support for science from Congress would both be considerably easier.

10. For the advanced reader: photographic emulsion and its reaction to light are described in G. Rieke, *Detection of Light from the Ultraviolet to the Submillimeter* (Cambridge: Cambridge University Press, 1994), 20.

11. H. Seliger, in the November 1995 *Physics Today* article on Roentgen, states that Roentgen's color blindness made him more discriminating of slight differences in the smallest quantities of light. Of all the people working on cathode-ray tubes in the late 1800s, Roentgen may have been the right person at the right time.

12. There is a collection of plates at the Harvard College Observatory that stretches back more than one hundred years. Other old observatories around the world have similar massive collections of plates. Handling these plates is an exercise in extreme caution. Not only are the glass plates fragile, but dropping one means you have perhaps wiped out an irreplaceable resource from the past, in addition to suffering the embarrassment of shattering glass.

13. Another drawback, the lack of linearity, will be presented in chapter 4.

14. Technical information on the Sloan Digital Sky Survey is available in D. York et al., *Astronomical Journal* 120 (2000): 1579, or online at <www.sdss.org>.

15. Our mesh is of much higher quality, however. Gold is often used because it is easily worked and conducts electricity well.

16. A more detailed description of proportional counters is available in the *McGraw-Hill Encyclopedia of Physics,* ed. S. Parker (New York: McGraw-Hill, 1983), and B. Ramsey, R. Austin, and R. Decher, *Space Science Reviews* 69 (1994): 139–204.

17. Particle-accelerator physicists obtain resolutions of a few millimeters, which are sufficient for their needs.

18. Much of the material describing the early years of X-ray astronomy was obtained from H. Bradt, T. Ohashi, and K. Pounds, *Annual Reviews of Astronomy and Astrophysics* 30 (1992): 391–427.

19. The June 1962 rocket flight was described in W. Tucker and R. Giacconi, *The X-ray Universe* (Cambridge, Mass.: Harvard University Press, 1985).

20. The height of 80 kilometers is an approximation. The height above which X rays are visible depends on the energy of the X rays. Higher-energy X rays penetrate deeper into the atmosphere before they are absorbed. These X rays can be detected at lower altitudes; the highest-energy X rays can be detected from high-altitude balloons.

21. This is a fine way of studying sources not visible to the eye, but it works only for objects that lie in the path of the Moon through the sky. In the early days of radio astronomy, when the detectors in that band did not provide high spatial resolution, the same technique was used to measure the size of radio-emitting objects.

22. The search-box plot is adapted from H. Gursky et al., *Astrophysical Journal* 146 (1966): 310.

23. For additional information about the relative levels of information across the electromagnetic spectrum, see Martin Harwit, *Cosmic Discovery* (Brighton, Eng.: Harvester Press, 1981).

24. The Orbiting Solar Observatory satellites included OSO-3, OSO-5, OSO-7, and OSO-8. Other spinners included the Astronomical Netherlands Satellite (ANS) and Ariel V.

25. These results are described in Bradt, Ohashi, and Pounds, *Annual Reviews of Astronomy and Astrophysics* 30 (1992).

26. The satellite was so named because it was launched on the anniversary of Kenya's independence.

27. This presumes that you, the looker, are not the driver. Another place to try this experiment is while riding a train or bus. You need to be relatively close to the objects you're passing, so planes and fast boats do not work so well: when you're moving quickly on a plane or boat, the scenery is usually relatively distant.

28. For the advanced reader: details may be found in P. Peebles, *Principles of Physical Cosmology* (Princeton: Princeton University Press, 1993), 646, and J. Peacock, *Cosmological Physics* (Cambridge: Cambridge University Press, 1999), 557.

29. A typical X-ray binary may emit 10^{38} ergs per second. If a mere 1,000 of the 10^7 to 10^{11} stars composing a typical galaxy are X-ray binaries, a galaxy will emit 10^{41} ergs per second.

30. The temperature units are degrees Kelvin (K stands for Kelvin), the absolute temperature scale in use in physics, astrophysics, and many other sciences. The zero point on the Kelvin scale is the temperature at which all thermal motion ceases. The zero point on the Celsius scale occurs at the freezing point of water. The relation between degrees K and degrees C (for Celsius) is a simple one: 0 degrees K is –273 degrees C. At 25 million degrees, who cares about the small difference between Celsius and Kelvin temperatures?

31. The argument essentially parallels the old joke about the person who, having lost keys on a dark street, searches for them under a streetlight because the light is better there. Given the choice of searching for exotic new particles of largely unknown properties that interact weakly with any detector or studying potentially new behaviors of particles with relatively well known properties, most reasonable people choose the well-lit path, at least until that path is eliminated as a possible explanation.

32. There are essentially two sets of reactions occurring at the center of the Sun: the proton-proton chain, which is the dominant energy generator, and the CNO tri-cycle. The proton-proton chain consists of three subcycles:

$$H^1 + H^1 \rightarrow D^2 + e^+ + v$$
$$D^2 + H^1 \rightarrow He^3 + \gamma$$
$$He^3 + He^3 \rightarrow He^4 + H^1 + H^1$$

$$He^3 + He^4 \rightarrow Be^7 + \gamma$$
$$Be^7 + e^- \rightarrow Li^7 + v$$
$$Li^7 + H^1 \rightarrow He^4 + He^4$$

$$Be^7 + H^1 \rightarrow B^8 + \gamma$$
$$B^8 \rightarrow Be^8 + e^+ + v$$
$$Be^8 \rightarrow 2\ He^4$$

where the symbols are: H = hydrogen, D = deuterium, He = helium, Be = beryllium, Li = Lithium, B = boron, e^+ = positron, γ = photon, e^- = electron, and v = neutrino.

The last subchain generates high-energy neutrinos but little energy. The lower the temperature at the core of the Sun, the more dominant the first chain becomes. The proton-proton chain requires a core temperature of at least approximately one to two million degrees to initiate nuclear reactions.

Treat these equations much as an accountant would: debits on one side of the arrow, credits on the other. By canceling quantities that appear on either side of the arrow and by noting that some of the reactions must occur twice for the next reaction to occur, you should see that the net result is:

$$4\ H^1 \rightarrow He^4 + Energy + 2e^+ + 2v$$

The atomic weight of hydrogen is 1.008172; helium has a weight of 4.003875. There is a difference of 0.0288 mass units that is converted to energy by that perhaps most famous equation, $E = mc^2$. Photons and neutrinos carry off the excess energy.

The CNO tri-cycle reactions are:

$$C^{12} + H^1 \rightarrow N^{13} + \gamma$$
$$N^{13} \rightarrow C^{13} + e^+ + \gamma$$
$$C^{13} + H^1 \rightarrow N^{14} + \gamma$$

$$N^{14} + H^1 \rightarrow O^{15} + \gamma$$
$$O^{15} \rightarrow N^{15} + e^+ + \nu$$
$$N^{15} + H^1 \rightarrow C^{12} + He^4$$

$$N^{15} + H^1 \rightarrow O^{16} + \gamma$$
$$O^{16} + H^1 \rightarrow F^{17} + \gamma$$
$$F^{17} \rightarrow O^{17} + e^+ + \nu$$
$$O^{17} + H^1 \rightarrow N^{14} + He^4$$

where C = carbon, N = nitrogen, O = oxygen, and F = fluorine. Again, the net result of this reaction is $4\,H^1 \rightarrow He^4 +$ energy $+ 2e^+ + \nu$. The CNO tri-cycle requires very high temperatures at the core, about 20 million degrees, before the reactions start to occur. A fuller description may be found in any book describing the structure and energy generation of stars, such as E. Bohm-Vitense, *Introduction to Stellar Astrophysics,* vol. 3 (New York: Cambridge University Press, 1992), 86–100.

33. J. Bahcall, the theorist, and R. Davis, Jr., the experimenter, recall the history of the solar neutrino experiment and predictions in Bahcall and Davis (1982).

34. The original neutrino observatory consisted of a large tank of one hundred thousand gallons of perchloroethylene, a cleaning fluid. Neutrinos interacted with atoms of chlorine from the fluid and changed the chlorine to argon. Argon does not react with other chemicals, so its presence can be detected relatively cleanly even at the level of a few atoms of argon in the hundred thousand gallons. Neutrino observatories must be located underground to shield them from other high-energy charged particles that could produce background events in the detectors.

35. The additional experiments include a detector in Japan called SuperKamioka (or Super K), a successor to the Kamioka detector of the mid- to late 1980s; SAGE, the Soviet-American Gallium Experiment in the Baksan Mountains of Russia (the "S" in the name predates the collapse of the USSR, but the experiment's name was drilled into everyone's mind); and SNO, the Sudbury Neutrino Observatory in Sudbury, Ontario. Each experiment has been designed to test specific neutrino reactions in the Sun using different materials for detection. SuperK and Kamioka both use water; SAGE uses gallium; SNO uses heavy water (D_2O).

36. G. Fiorentini et al., in a 2001 preprint (available at the physics preprint archive at <xxx.lanl.gov/form/> as preprint hep-ph/0109275), derive the temperature, assuming sterile neutrinos do not exist. Should sterile neutrinos exist, the temperature would become a minimum temperature.

37. The first use of MACHO in print appears to be B. Carr and J. Primack, *Nature* 345 (1990): 478, which attributes the acronym to K. Griest of the University of California at Berkeley.

38. For the advanced reader: see B. Oppenheimer et al., *Science* 292 (2001): 698.

39. See, for example, J. Binney, and M. Merrifield, *Galactic Astronomy* (Princeton: Princeton University Press, 1998).

40. Forty-nine square degrees is the area on the sky occupied by approximately 200 full moons.

41. There is no "correct" number because the total depends on how faint are the objects one observes. However, at a given intensity level, one does have a "correct" answer that ROSAT supplied in the early 1990s: about 55,000 to 60,000 objects to the level of the Einstein survey.

42. The relationship is: $L = 4\pi R^2 \sigma T^4$, where σ is the Stefan-Boltzmann constant and the other quantities are defined as luminosity, L; size of radius, R; and temperature, T.

43. An erg is a unit of energy; ergs per second represent power or luminosity. The use of erg as a unit of energy is typical in astrophysics, but power units are more familiarly known as watts (named for the Scottish inventor and engineer James Watt). The solar luminosity is about 2×10^{26} watts.

44. Details are available in D. Schwartz et al., *Astrophysical Journal (Letters)* 540 (2000): 69–72. The X-ray luminosity is $\sim4\times10^{44}$ ergs/second, where the luminosity of the sun is $\sim3.8\times10^{33}$ ergs/second.

45. The first explanation of a projection effect appears to be M. Rees, *Nature* 211 (1966): 468. The apparent speed of transverse motion is related to the actual speed by $\beta_{obs} = \beta \times \sin\theta / (1 - \beta\cos\theta)$, where β_{obs} is the observed speed of motion in units of the speed of light, β is the actual speed, and θ is the angle between the line connecting the observer and the target, and the target and the jet. From this relation, one cannot measure β and θ separately, but can place limits on them.

46. Astronomers constructed other models as alternatives to the invocation of a galactic-mass black hole. A highly concentrated cluster of stars was one alternative. As described in J. Frank, A. King, and D. Raine, *Accretion Power in Astrophysics,* 2d ed. (New York: Cambridge University Press, 1992), the high concentration of stars would not provide power for the length of time needed to describe active galaxies.

47. $r = 2\,GM/c^2$, where G is the gravitational constant and c is the speed of light.

48. Recently, researchers have accumulated evidence of "intermediate mass" black holes, objects with masses about 1,000 times the Sun's mass. The jury is still out considering the nature of these objects.

49. At least, not on state police cars in Massachusetts. Strobes appear to be replacing the top-mounted "bubble-gum machine" lights, presumably because they are less expensive to maintain and operate.

50. A light-year is the distance traveled by light during one year; a light-day is the distance traveled by light during one day. Equivalent definitions apply for light-hour, light-minute, and light-second. The light-day is about 173 AU = 2.6×10^{10} kilometers = 1.6×10^{10} miles. One AU equals one astronomical unit, the average distance between Earth and the Sun, and corresponds to about 149 million kilometers = 93 million miles. The light-hour is about 7 AU = 1.1 billion kilometers = 6.8×10^8 miles; the light-minute is about 18 million kilometers = 11 million miles; the light-second is 299,792 kilometers = 188,000 miles, which is less than the distance from Earth to the Moon.

Chapter 3

1. This, apparently, is how one ends up with schools of art.

2. Seurat's contribution is described in G. Pischel, *A World History of Art,* 2d rev. ed. (New York: Newsweek Books, 1978). The phrases "nebulous illusions" and "scientifically" come from this text.

3. Actually, Seurat went even further: instead of placing green dots where a leaf might be, he placed dots of blue and yellow. When the painting is viewed from a few feet away, the eye and brain combine the dots into green.

4. Neither is your eye a perfect detector, for that matter.

5. From the point directly overhead to the horizon is 90 degrees. Hold a (United States) penny at arm's length with the Lincoln Memorial facing you. One degree, at arm's length, is about the width of the Lincoln Memorial, to better than 10 percent accuracy. There are 12 columns depicted in that image. Because 1 degree = 60 arc-minutes = 12 times 5, the distance between two of the columns is about five minutes of arc. One minute of arc is then the distance between two of the columns divided by five. It is about seven thousandths (0.007) of an inch, or 0.18 millimeters, at arm's length.

6. This unit was named for the German physicist Heinrich Hertz (1857–94) and has nothing to do with a particular car-rental company.

7. Radio astronomy started in 1931 with the radio engineer Karl Jansky, who worked for Bell Telephone Laboratories. He studied the effects of thunderstorms on radio telephone circuits. In doing so, he uncovered a "steady hiss type of static" that moved across the sky. Subsequent work established that the signal originated in our galaxy, the Milky Way. Another radio engineer, Grote Reber, followed up on Jansky's work by building his own radio telescope in the late 1930s. His first map of the radio sky attracted attention from several groups around the world, but progress languished until World War II ended. Radio astronomy developed quickly during the 1950s. (Summary of a historical section in J. Krauss, *Radio Astronomy* [New York: McGraw-Hill, 1966].)

8. The "or" means that the two ways of writing these numbers are equivalent. I trust the reader sees the value of the shorter notation: for numbers typically tossed around in astrophysics, we save paper. For those unfamiliar with scientific notation, a brief review follows. A positive exponent (the "6") represents the number of zeroes to the right of the leading digit; a negative exponent represents zeroes to the left of the trailing digit. Zeroes in front of a number must include a decimal, so it describes a fraction. Consequently, positive exponents represent very large numbers compactly; negative exponents, very small fractions.

9. The estimate for the horizon size comes from standard navigation books on the dip of the horizon, the amount by which one corrects the altitude of a star for the observer's height above sea level.

10. For the advanced reader: $E = h\nu$, where h is Planck's constant, 6.63×10^{-27} ergs per second or 4.14×10^{-15} keV per second.

11. For the fundamentally antimetric crowd: $c = 1.1803 \times 10^{10}$ inches per second = 9.8357×10^{8} feet per second = 186,280 miles per second = 6.706×10^{8} miles per hour.

12. I cite these numbers as a range because the exact value is difficult to find. Dr. P. Gouras states "a million different colors" in a medical discussion of vision: <webvision.med.utah.edu/Color.html>. M. Abrash, writing in a computer journal in 1992, cited experiments conducted at the Jet Propulsion Laboratory in the 1970s that revealed people who could distinguish as many as 50 million different colors in video displays. The number clearly depends on individual response as well as on the conditions under which the test is conducted. See M. Abrash, *Dr. Dobb's Journal* 17, no. 8 (1992): 149.

13. These numbers come from the third edition of the EGRET catalog. See R. Hartmann et al., *Astrophysical Journal Supplement* 123 (1999): 79.

14. At least it should have vanished. For the advanced reader: the results were reported by R. Sambruna et al., *Astrophysical Journal* 538 (2000): 127.

15. In the original draft text of this section, I mentioned that the longest radio waves discovered were those with wavelengths of about a meter. One of the referees of that draft corrected me by noting that Schumann resonances had been measured to three to eight hertz, frequencies corresponding to wavelengths of some one hundred thousand kilometers. That referee also provided a reference: J. D. Jackson, *Classical Electrodynamics*, 2d ed. (New York: John Wiley and Sons, 1975), 364. I found this fascinating because, although I used that text in graduate school, our class skipped that chapter, and I was likely too busy to read an extra chapter of Jackson's book, a book infamous among graduate students for its difficult end-of-chapter problems. Schumann resonances, it turns out, use the entire ionosphere of Earth as a resonant cavity.

16. Shortly after the launch of the Hubble Space Telescope, scientists discovered its mirror defect. The defect turned out to be an incorrect shaping of the mirror, by about one millimeter, which left it with a curvature that would not focus all the light to a single point. In other words, the point-spread function of the telescope was considerably poorer than the design specified. The design called for 70 percent of the light to fall into a narrow core; the on-orbit measurement showed that about 15 percent of the light did so. The Hubble Space Telescope was built to provide observations with high spatial resolution and observations in the ultraviolet band.

The incorrect shape was bad enough, but the press unfortunately muddied the situation. After scientists discovered the mirror flaw, at least one major news organization dramatically reported that the telescope "did not work." This created the inaccurate impression in the public's mind that Hubble was a piece of expensive, useless junk orbiting Earth. Quite a few scientists used the telescope for about two and a half years before the first repair mission in 1993. High-spatial-resolution imaging was definitely difficult to obtain. Images in the ultraviolet could still be obtained, however, and represented significant science, even though the results were less than those expected from the original design, simply because ultraviolet images were few in number. The press chose not to note the differences, confusing the public in the process.

One of the early referees for this book adopted the position that Hubble's "only advantage" over ground-based telescopes is the high spatial resolution, so in this referee's opinion, the press had the story correct. However, the author and many other scientists found the ultraviolet capability highly valuable. Furthermore, adaptive optics systems, attached to all large modern telescopes, actively compensate for the atmosphere and thereby encroach on the high-spatial-resolution domain that seemingly is Hubble's primary advantage. Finally, the author remembers quite clearly answering many questions from fellow taxpayers during the two years preceding the repair mission whenever Hubble science results were presented. Each question, clearly demonstrating that people were not only interested but also thought about the material they read, started in a similar manner: "If it [Hubble] does not work, how did they get the data?"

17. Recall that the lower-quality point-spread function for the Einstein mirrors compared to the Chandra mirrors was a conscious choice because Einstein represented the first imaging X-ray observatory.

18. I intentionally use the word "deflect" rather than "reflect" because the former is more general. X-ray mirrors do reflect, but only at a shallow grazing angle. "Deflect" should help the reader keep this in mind.

19. To quote *The Simpsons'* Homer, "D'Oh!"

20. The 4,697th entry in the *New General Catalog of Nebulae and Clusters* by J. Dreyer and R. Sinnott, a catalog of nonstellar objects.

21. The details are available in C. Sarazin et al., *Astrophysical Journal (Letters)* 544 (2001): L101.

22. The details for NGC 1553 are available in E. Blanton et al., *Astrophysical Journal* 552 (2001): 106.

23. This model is the work of L. Ciotti and co-workers and is described in L. Ciotti et al., *Astrophysical Journal* 376 (1991): 380.

24. For the reader unfamiliar with this map, here is a brief explanation. To flatten the surface of the globe to match color figures 3 and 4, picture yourself standing inside a transparent globe. Make a cut along the International Date Line. That cut becomes the outer edge of the image. The vertical line in the middle of the image is the line of zero longitude, otherwise known as the Greenwich meridian. Color figures 3 and 4 are similar, except that the vertical line of zero longitude cuts through the center of our galaxy and its poles (just as the zero-longitude line does on the globe). Galactic coordinates place the center of our galaxy at the center of the map. Because our galaxy resembles a flattened disk, our position inside the disk means that we see a thin band on the sky, the galactic plane. Sources that lie in the disk of our galaxy will then be found to lie in that plane. The center of the galaxy lies in the direction of the constellation Scorpius. Readers who have access to dark skies and live near the equator can easily see, at particular times of the year, the flat disk plus the spherical nucleus that make up the structure of our galaxy.

25. Observations defining the properties of the local hot bubble largely came from the efforts of astrophysicists at the University of Wisconsin. See D. McCammon and W. Sanders, *Annual Reviews of Astronomy and Astrophysics* 28 (1990): 657.

26. For advanced readers: J. Maíz-Appellániz, *Astrophysical Journal (Letters)* 560 (2001): L83, describes the result.

27. One direction along which the intervening absorption is low is known as "Lockman's Hole" after F. Lockman, a radio astronomer at the National Radio Astronomy Observatory. He has never been overly happy with that name, as he told me in a private conversa-

tion some years ago. The name is essentially ensconced in the literature; polite authors and editors, citing propriety, have altered the name slightly to "Lockman Hole."

28. The Chandra teams reported results in R. Mushotzky et al., *Nature* 404 (2000): 459, in R. Giacconi et al., *Astrophysical Journal* 551 (2000): 624, in G. Garmire et al., *Bulletin of the American Astronomical Society* 32 (2000): abstract 261, and in P. Tozzi et al., *Astrophysical Journal* 562 (2001): 42. The Newton results were presented in G. Hasinger et al., *Astronomy and Astrophysics* 365 (2001): L45.

29. As reported in R. Giacconi et al., *Astrophysical Journal* 551 (2001): 624. The R. Giacconi who is the first author of this paper is the same R. Giacconi who located the first X-ray source outside the solar system (Sco X-1), discussed in chapter 2.

30. The use of the word "deep" may be confusing: it denotes a probe to very faint light levels, or, equivalently, a probe outward to very distant objects.

31. This date comes from a timeline of CCDs on the Sensors page at NASA-HQ: <www.photocourse.com/01/01-04.htm> or <zebu.uoregon.edu/ccd.html>.

32. CMOS stands for Complementary Metal Oxide Semiconductor, also known as "artificial retina" because the technology changes the problem of imaging from one of devices to one of circuits (to paraphrase a boxed comment at the IEEE (Institute of Electrical and Electronics Engineers) Solid State Circuits Technology Workshop on CMOS Technology Web site). CMOS imagers are based essentially on the same technology as standard computer memory chips. Their development lagged because the technology and the problem of image sensing appeared incompatible. Early efforts were plagued by enhanced noise and temporal instabilities. Researchers at NASA's Jet Propulsion Laboratory eliminated the roadblocks in the early 1990s by finding a way to make CMOS imagers with nearly scientific quality.

33. CCDs were first attached to optical telescopes in the mid-1970s, and their use was widespread by the early 1980s.

34. ASCA stands for Advanced Satellite for Cosmology and Astrophysics. In Japan, the engineers who launch the satellite choose the name. The actual name, Asuka, means "Flying Bird" in Japanese. It is pronounced "asca," and is apparently a pun.

35. This is not true of the vast majority of camcorders currently on the market: the output of a CCD in a camcorder is written to a tape (analog) and then played back through a VCR. The next generation of camcorders will be digital; some had already appeared by late 1999. The players (e.g., DVD) must also be digital. Still cameras have already made the transition, with increasing numbers recording directly to a CD-ROM. Software such as GIMP (Linux) allows the user to alter a poor-quality image.

36. G. Amelio first presented this analogy in an article in *Scientific American,* February 1974, 23.

37. Following this long description, you might wonder why a particular bucket is not just emptied and measured. CCDs fundamentally must use the row-by-row transfer; in other words, each pixel does not have an individual address. CIDs (Charge Injection Devices, a form of CMOS imager) do have pixel-by-pixel addressing but were never developed because of noise and temporal-instability properties. The impact on science can be important: with individual pixel addressing, the "frame rate," the speed with which every pixel can be read, can be higher, so fewer X rays are missed. The number of missed X rays is estimated by counting the length of time during which the chip is being read by the electronics instead of observing the sky. This time length is called "dead time," the amount of time the instrument is incapable of collecting science data.

38. The first X-ray telescope flown using solid-state detectors was the shuttle-borne Broad Band X-ray Telescope, an experiment assembled by the X-ray Astrophysics Group at the NASA-Goddard Space Flight Center. A shuttle carried it aloft in December 1990. The detectors were lithium-drifted silicon semiconductors, and their spatial resolution was poor. At the time, the detectors had the best spectral resolution and, together with the gold-coated foil mirrors, served as a proof of concept for ASCA. The detectors were CCDs, however.

39. Although the taxpayer receives benefits, such benefits are not the prime reason to spend taxpayer dollars to carry out research in astrophysics. See chapter 9 for a discussion of the prime reason. The development of CCDs and CMOS imagers was at least partially funded by taxpayer money. Both technologies have led to commercial developments, jobs, and a sizable return on the taxpayers' investment.

Chapter 4

1. C. Hayashi, *Annual Reviews of Astronomy and Astrophysics* 4 (1966): 171.

2. The orbit of Pluto has a mean radius of about 29 AU; the actual size of the solar system is larger. Two Pioneer and two Voyager spacecrafts, originally used to provide the first images of the outer planets, are continuing their journeys outward. Scientists hope they will cross the boundary from solar system to non-solar system soon. The boundary will essentially define the outer edge of the solar system.

3. The terms "soft" and "hard" are not rigid, but generally depend on the detector used for the study. One can define the lower-energy photons as soft for any bandpass. Red photons in the optical band can be described as soft optical photons, relative to the harder photons of blue light. Convention in the X-ray bandpass considers soft X rays to have energies less than about 1 keV, but this is arbitrary.

4. A detector with an efficiency of zero would, presumably, remain in the lab for additional work. One of the goals of every instrument builder is increased efficiency.

5. The exact distance does not matter, so long as one of the objects is twice as far away as the other and both of them are not nearby. You can use lights a hundred yards and two hundred yards apart (in other words, one soccer field away and two fields away).

6. A law of physics is a behavior pattern repeatedly observed. These laws are actually behaviors that have been established so well that we can rely on them. Consequently, despite the commentary of sports announcers, no one ever defies the laws of physics.

7. The story of the discovery of quasars is interesting on its own. In the early 1960s, radio astronomy had identified a number of starlike objects. Spectra of these objects showed a puzzling set of emission lines that defied easy identification. In 1963, M. Schmidt at the Palomar Observatory in California realized that the lines could be identified as lines of hydrogen if the object was Doppler-shifting away from us at about 15 percent of the speed of light. Because the recession velocity is proportional to the distance (the relationship discovered by E. Hubble that signals the expansion of the universe), quasars had to lie at what was, for 1963, an enormous distance. Since that time, quasars have been found receding with speeds over 90 percent of the speed of light. The most distant quasar so far discovered has a redshift of 6.28 (SDSSp J103027.10+052455.0), reported in R. Becker et al.(2000); no doubt this record will fall any week now. Additional information is available in nearly any introductory astronomy book, such as G. Abell, *Exploration of the Universe*, 4th ed. (Philadelphia: Saunders College Publishing, 1982).

8. The reader should note that the existence of black holes remains undemonstrated. We infer from observations that an object exists that behaves like a black hole. As professionals, we act as if black holes exist. A future X-ray satellite may actually obtain an image of a black hole.

9. Specifically, the X-ray brightness is about 10^{-4} of the total luminosity ($L_x \sim 10^{-4} L_{tot}$), where L_{tot} includes the emission from all wavelengths.

10. For the advanced reader: this is known as Hooke's Law in physics, $F = -k\,x$, where k is the force constant and x is the amount of compression of the spring. "k" differs for each spring. More-modern scales use "load cells," which convert a mechanical deviation such as a load (weight) into a change in electrical properties. The change is directly proportional to the deviation, so it provides an excellent measure of the weight. There are no moving parts, so the scale lasts longer.

11. The instrument teams were composed of people from the Pennsylvania State University, the Massachusetts Institute of Technology, the Smithsonian Astrophysical Observatory, and members of the X-ray group at NASA-Marshall Space Flight Center.

12. The tests we pursued at XRCF included those for count-rate linearity, molecular contamination, point-spread function (core, wings), and effective area.

The count-rate linearity test steps the beam intensity along a predetermined set of changes in intensity. The test checks that for every known number of photons entering the mirror-detector combination, a constant fraction of photons leaves the detectors. As the number of entering photons increases, the number of outgoing photons should also increase. This test is then the precise optical equivalent of placing light and heavy weights sequentially on a bathroom scale. The scale should show in linear terms the difference in the weights.

Molecular contamination can be fatal for X rays. X rays scatter off the smallest quantities of dust; X rays are also absorbed by the smallest quantity of molecules on the mirrors. The molecular-contamination test is designed to measure the number of molecules on the surface of the mirrors. If we see a decrease in the detected X rays at particular energies where the molecules are known to absorb energy, then we know we have a dirty mirror and we must clean it. (Perhaps that's why we exist in this universe: to clean up the dirt.)

The point-spread function test checks the quality of the optics as described in chapter 3. The effective-area test measures the overall sensitivity of the telescope and the detectors.

13. Too much air attenuates an X-ray beam to the point of uselessness. This is the reason that, when you are sitting in the chair in a dentist's office, the dentist places the X-ray generator (that [usually] black thing on the swivel arm) directly against your cheek. To obtain a clear set of X rays of your teeth, the beam must be strong, but not so strong that the X rays damage the cells of your cheek, teeth, or mouth. The only way the dentist could use a lower dose would be to pump the air out of the room (or, at least, the air between your head and the X-ray gun). The most important part of the support machinery at XRCF was the vacuum pump, basically a well-built vacuum cleaner designed to remove the air from the chamber. Without the vacuum pumps, the Calibration Facility would not work.

14. I was part of the XRCF test team in February and March 1997. We worked 12-hour shifts that stretched to 14-hour days, at a minimum, for two solid weeks. During that time, we had two tornado warnings, both at about 2:00 A.M. (for those who do not live in tornado country, a warning means a tornado has been sighted), and one outbreak of smoke in the Source Building. The cause of the smoke was never determined. I have new respect for the firefighters who, carrying fire extinguishers, charged into a smoky building containing hundreds of high-voltage cables.

15. For advanced readers: for an ellipse, the speed is the square root of the quantity $GM[(2/r) - (1/a)]$, where G is the gravitational constant, M is the mass of the central object (for example, the Sun), r is the distance from the central object to the orbiting object, and a is half the length of the major axis of the ellipse. The relationship also applies to a hyperbola if the minus sign is changed to a plus sign. The interested reader may consult any book on orbit dynamics.

16. The term "dirty snowball" or "dirty iceberg," first used by Dr. F. Whipple in 1950, remains the best description of the model for comets. Comets are composed of frozen gases (the "snowball") and dust (the "dirty" part). The gases include a long list of organic compounds. See, for example, C. Sagan and A. Druyan, *Comet* (New York: Pocket Books, 1985).

17. Hyakutake is pronounced "Yah-ku-ta-key," without an accent on any syllable.

18. The passage of comet Hyakutake in 1996 is a perfect example. As a professional, I'm used to looking at things in the sky that basically have not moved appreciably in my lifetime. I thought Hyakutake was spooky because, using just a pair of binoculars, I could *see* it moving.

19. This argument applies to all wavelengths, not just to the X-ray band. In the optical band, however, objects are usually sufficiently bright and sufficiently pointlike that they are visible. Very small objects do elude detection; as a result, people have pushed recently for telescopes to survey the sky looking for small, as yet undetected asteroids. At least two such projects are in operation: Project LINEAR (see note 22) and Project NEAT (Near Earth Asteroid Tracking). The NEAT team publicized its discovery that the number of nearby asteroids had been overestimated by about a factor of two. Previous estimates placed

the number of asteroids larger than one kilometer in the low thousands; NEAT data suggest there are about five hundred to a thousand such asteroids. (D. Rabinowitz et al., *Nature* 403 [2000]: 165.)

20. This will not be true once a particular type of observatory, an interferometer, is in routine use. An X-ray interferometer may fly sometime before 2020. Ground-based optical interferometers are coming into use now; a satellite optical interferometer should fly before 2010.

21. The description of the X-ray emission of comets, particularly Hyakutake, is based on a brief science summary in *Physics Today* 50, no. 12 (December 1997), by C. Day.

22. LINEAR stands for LINcoln Near Earth Asteroid Research (program), where the Lincoln is the Massachusetts Institute of Technology's Lincoln Laboratories, Bedford, Massachusetts. The research program has as a goal the inventory of near-Earth asteroids by repeatedly imaging the sky to search for anything that moves. Detection of asteroids reaches down to magnitudes of about 20; typical asteroids have an albedo, the number that describes the percentage of light reflected from the surface, of about 0.10. The magnitude limit means that objects down to about a kilometer in size can be detected. The NEAT project is a scientific competitor. The detection of comets comes along for free.

23. The greater the number of comets moving around in the inner solar system, the greater the chance of a disaster caused by a collision. I'd prefer to see an inspired astrophysicist come along. One comet impact per solar system in our lifetime is enough.

24. Astronomers have known for a long time that the numbers did not add up: there were too few supernovae per year to generate all the gold, iron, oxygen, and so forth that we see. These elements were not created during the early phases of the universe (the Big Bang) because it expanded and cooled too quickly for nuclear reactions to commence. Recent computer simulations suggest that the merger of two neutron stars, the corpses of stars that have exploded, may have produced the remainder. The simulations were carried out by Dr. Stephan Rosswog and his coworkers at the Universities of Leicester (United Kingdom) and Basel (Switzerland). A report is available at <http://www.ukaff.ac.uk/pressreleases/release2.shtml>; it was released to the press on March 29, 2001.

25. Additional mechanisms exist. The shock that tears apart the star will also produce a thermal burst as it breaks through the outer layers of the star. This burst may briefly be hot enough to produce X rays. Inverse Compton scattering, the collision of a photon with an electron in which the photon gains energy, may also explain some X-ray emissions.

26. Compton scattering was named for the American physicist Arthur Compton, who discovered it in 1922; he won a Nobel Prize in physics for that discovery.

27. The dozen or so include SN1987A, the closest supernova in more than three hundred years. It lies in the Large Magellanic Cloud, a satellite galaxy of our own. SN1987A emitted X rays by both mechanisms described: Compton scattering of gamma rays from the decay of radioactive matter and the interaction of the outgoing shock with the circumstellar medium. To date, it remains unique.

28. All pulsars are neutron stars, but not all neutron stars become pulsars. We do not know why.

Chapter 5

1. Information, that is, other than the fundamental information about a footrace: the winner.

2. Typical 100-meter Olympic or world record times are in the 9- to 10-second range. The average speed to move 100 meters is then 10 to 11 meters per second. Because a meter is slightly more than 3 feet, the average speed is equivalent to about 31 feet per second.

3. Lucky for us: life on Earth would be very different if the Sun were a variable source over short spans of time.

4. Pulsating sources, objects that "breathe," can also vary in a repetitive pattern. For now, let's ignore this complication.

5. "Many" here is a relative term, as I did not count the numbers of satellites in low- and high-Earth orbits: communications satellites are generally in high-Earth orbit, sufficiently high that they orbit at the same rate that Earth turns. They remain over the same spot on Earth, which is a major convenience for communications because, with a minimum of three or four satellites, complete planetary coverage can be achieved, as was argued by Arthur C. Clarke (*Wireless World,* October 1945). Weather satellites are also parked in high orbit. Military imaging satellites may be in low orbits, the better to get good ground resolution. Astronomy satellites are usually in low orbit because of the expense; the *International Ultraviolet Explorer,* a joint NASA-Europe spectroscopy mission, was placed in a high orbit.

6. The Rossi X-ray Timing Explorer and the Hubble Space Telescope both orbit at about 550 to 600 kilometers of altitude. Hubble must be in a relatively low orbit so that it can be serviced by the space shuttle.

7. Although the Dow Jones Industrial Average is not interrupted from day to day, financial transactions do occur elsewhere on the globe while the market in New York sleeps. If some of the transactions cause substantial financial changes, then when the market does open, the DJIA will immediately react to that financial news. Essentially, the forced time offline inserts a lag between the time investors hear the financial news and the time they can act on that news. Investors will gradually push all financial markets into 24-hour trading, at least during the business week, to eliminate information lags and to gain the freedom to react to financial events whenever they occur. In June 1999, the New York Stock Exchange announced that it intended to explore extended trading hours.

8. EXOSAT was the first dedicated X-ray observatory placed into an elliptical orbit. The orbital period was about 90 hours with a perigee of about 350 kilometers and an apogee of about 191,000 kilometers.

9. As noted earlier, in the discussion of neutrinos, the exact method by which a star is blown apart is still the subject of vigorous research.

10. For anyone using syringes, this number should seem appropriate. For the rest of us, it presents a perfect opportunity to illustrate changes in units. Any decent measurements book contains volume equivalences; one (J. Tuma, *Handbook of Physical Calculations,* 2nd ed. [New York: McGraw-Hill, 1983]) lists that one liter is 10^{-3} cubic meters (or 10^{-3} m^3). Because one hundred centimeters equals a meter, the arithmetic is as follows: One liter = 10^{-3} m$^3 \times (10^2$ cm / 1 m$)^3 = 10^3$ cm^3. One milliliter = 10^{-3} liter = 1 cm^3. Five cm^3 = 5 milliliters; the volume a teaspoon holds is about 5 milliliters.

11. For the advanced reader: see S. Das Gupta and G. Westfall, "Probing Dense Nuclear Matter in the Laboratory," *Physics Today* (May 1993): 34.

12. This is quite true, based on the calculations of F. Adams and G. Laughlin, *Reviews of Modern Physics* 69 (1997): 337.

13. The neutron-star equation of state dictates how rapidly the matter cools. The equation of state relates the three quantities of pressure, temperature, and density of the matter. The ideal-gas law, $P = nkT$, where P = pressure, n = number density (particles/cm^3), k = Boltzmann's constant, and T = temperature, is frequently used to describe gases in certain conditions and is an example equation of state. The equation of state for neutron-star matter has been a subject of research for decades. Once determined, the equation can be used to predict how rapidly neutron stars cool, among many other things.

14. The isolated neutron star, RX J185635-3754, is described in F. Walter and L. Matthews, *Nature* 389 (1997): 358.

15. The gravitational influence extends to infinity. However, gravity is an inverse square force ($1/r^2$), so there is a distance beyond which the gravitational pull of a particular object becomes so small as to be negligible.

16. Or, without any implied commercial endorsement, two Hershey Kisses kissing.

17. Many assumptions are made in defining the Roche critical surface in this manner. Astronomers opposed to the concept of the critical surface point out the possible ways to fail to achieve the simplistic Roche critical surface as described. For example, no accelerations other than gravity and centrifugal must be present. Pressure forces within the individual stars may matter; so, too, may radiation forces from one star on the other. The

masses of the two stars are treated as if they were point particles. The complete problem is mathematically complex, not least because it requires three dimensions for a proper simulation. However, we observe interacting binary stars that act as if the simple critical surface as described applies to them. The problem is fundamentally one of hydrodynamics; it will be solved once the computing power exists.

18. The argument assumes that Earth is a point source, while the Moon is treated as two spheres of half the mass. The problem then becomes at what radius the gravitational attraction toward Earth of the near sphere exceeds its gravitational attraction to its twin. This radius is the critical Roche limit or the differential tidal limit.

19. The calculations involved in understanding Roche-lobe overflow are horrendously complex. The problem is fundamentally three-dimensional in nature, so few or no simplifying assumptions are possible. Low-mass stars also undergo convection in their outer layers; the upper layers near the overflow point will be turbulent. (Statements based on R. Gilliland, *Astrophysical Journal* 292 [1985]: 522–34.)

20. Technically, stars rotate, or spin, on their axes and revolve, or orbit, about each other.

21. The visibility of a Nobel Prize probably attracted the attention of entrants to NASA's essay contest to suggest an observatory name for what, at that time, was known as the Advanced X-ray Astrophysics Facility. "Chandrasekhar" was the winning entry, submitted by Tyrel Johnson, then a student at Priest River Lamanna High School in Priest River, Idaho, and Jatila van der Veen, a physics and astronomy teacher at Adolfo Camarillo High School in California. That's how AXAF became known as the Chandra X-ray Observatory.

22. The numbers are approximate. For the white dwarf, the number is approximately 1.5 solar masses because the exact value depends on the composition of the white dwarf. For the neutron star, the range of 1.5 to 3 solar masses is approximate because we do not know the equation of state for matter at densities believed to be typical of neutron stars. The value for black holes is even more uncertain, but essentially depends on the upper limit for the neutron star. Note that a lower bound of about 3 solar masses should not imply that black holes of lower mass cannot exist. It simply means that (i) stellar cores have a difficult time shedding excess mass; (ii) a collapsing core can be supported up to about 3 solar masses; and (iii) beyond that mass, no support appears possible. Black holes have not been demonstrated to exist, but the circumstantial evidence is accumulating. It is possible that advances in a quantum theory of gravity will find some new state of matter that permits collapsing cores to survive to higher masses.

23. The upper bound on black holes evolved from stellar cores must be lower than the maximum mass a star may have, because during a star's evolution, it naturally loses mass. Mass loss is as yet a rather poorly understood aspect of stellar evolution.

24. I use the word "candidate" because of the comment earlier about black holes having not been proved to exist. Many astrophysicists act as if black holes have been demonstrated to exist because the circumstantial evidence has reached the stage of "compelling."

25. Described in J. McClintock, *Accretion Processes in Astrophysical Systems,* ed. S. Holt and T. Kallman (New York: American Institute of Physics, 1998), 290.

26. If astrophysicists had to conjure up unique names for every star or galaxy detected, we'd never do anything else. The easier approach uses as the name the coordinates that describe the location of the object in the sky. If we used this approach for our cities and towns, then New York would be called "W74.0, N40.75," a name that would uniquely identify the city. "1655-40" tells us that the object was discovered at right ascension 16 hours 55 minutes, declination –40 degrees; the "J" is the designator for the epoch beginning in 2000. The "GRO" portion identifies the satellite or catalog that discovered or first listed the object, in this case, the (Compton) Gamma-ray Observatory.

27. The GRO J1655-40 mass estimates are from C. Bailyn, et al., *Nature* 378 (1995): 157.

28. This material is based on the translations and analyses of F. Stephenson ("East Asian Observations," in *Applied Historical Astronomy,* Joint Discussion 6, Twenty-fourth meeting of the International Astronomical Union, Manchester, England, August 2000), and I. Hasegawa (*Vistas in Astronomy* 24 [1980]: 59–102). Both authors compiled records from

China, Korea, and Japan starting as early as 1300–1500 B.C. The records include observations of novae, supernovae, comets, aurorae, meteors, and solar eclipses.

29. The rate is listed in Y. Tanaka and N. Shibazaki, *Annual Reviews of Astronomy and Astrophysics* 34 (1996): 639.

30. A second explanation, first proposed in 1969 (B. Paczynski, J. Ziolkowski, and A. Zytkow, in *Mass Loss from Stars,* ed. M. Hack [Dordrecht: Reidel, 1969]), remains unsupported by data: an instability in the red companion's rate of mass loss. This explanation predicts that the disk will become bright at the outer rim, which has never been observed.

31. An increase in the viscosity of the disk increases the rate of accretion. Viscosity in accretion disks is little understood and remains a topic of vigorous research.

32. The naming scheme for variable stars is as follows. The first variable star in a constellation is called "R Con," where "Con" is the Latin name for the constellation. The second variable is "S Con," the third "T Con," and so on down to Z. The list then starts at "RR," "RS," "RT," down to "ZZ." It starts over at "AA," "AB," to "QZ." That gives 334 variable stars per constellation. After that, each new variable is given the designation "Vxxxx," where the "xxxx" is a number that starts at 335. SS Cygni is thus the nineteenth variable star in Cygnus. This is another of those interesting, historical, quirky forms of nomenclature in astronomy.

33. From the catalog of J. van Paradijs, *X-ray Binaries,* ed. W. H. G. Lewin, J. van Paradijs, and E. P. J. van den Heuvel (Cambridge: Cambridge University Press, 1995), 536.

34. The satellite is named for Bruno Rossi (1905–93), an astrophysicist from the Massachusetts Institute of Technology, who pioneered many of the early X-ray observations in the 1960s and 1970s. Of the sources shown in Figure 5.5, the light curves of Aql X-1, XTE J1550-564, GX 339-4, 1705-440, 2012+38, Cen X-3, and 1942+27 were all obtained with the ASM.

35. Don't laugh, but the "fast" chips onboard RXTE are 286s. At the time RXTE was designed, these were the chips of choice that had been demonstrated to be tolerant of the harsh radiation environment of low-Earth orbit.

36. Earth, you say? Earth emits X rays for the same reason that comets are X-ray sources: the charged particles in the solar wind hit the oxygen and nitrogen atoms and molecules in the upper atmosphere. Emission from X rays in the 0.4–0.6-keV band occurs via charge exchange. See note 39.

37. The road map is available at the Office of Space Science Web page at NASA <http://spacescience.nasa.gov>.

38. The original paper is K. Prendergast and E. Spiegel, *Comments on Astrophysics and Space Physics* 5, no. 2 (1973): 43.

39. Earth's aurorae are produced when charged particles from the Sun, funneled to the polar regions by the Earth's magnetic field, crash into oxygen and nitrogen atoms and molecules high in the atmosphere. The basic electrochemical reactions are: $N_2 + e^- \rightarrow N_2^+ + e^- + e^-$ and $O + e^- \rightarrow O$ (excited) $+ e^-$. This is the reaction described in note 36 that causes Earth to emit X rays.

40. As noted by W. Lewin, in *The Hot Universe,* ed. K. Koyama, S. Kitamoto, and M. Itoh (Dordrecht: Kluwer Academic Publishing, 1998), 111.

41. The ASM XTE movie is available at <http://xte.mit.edu/XTE/movie/>. At least two versions exist for downloading, one a short six-month sample as well as the full three-year view.

Chapter 6

1. Unfortunately, I no longer have the name of the astrophysicist who first stated this. Whoever it is should have his or her name entered in the pantheon so that we all learn the quotation at a young age.

2. Look at any photograph from a recent family outing. How do you know the height of the people in the photo? If you answer "because Jimmy (or Susie) is five feet tall," that's cheating. If you had placed a board, marked off in 1-foot increments, next to the people when you photographed them, their heights would be calibrated.

3. Politicians are not the only group to use polls, in spite of the impression conveyed in the press and broadcast media. Advertisers and the media themselves use polls to find out what people are willing to buy or to watch. Both groups respond accordingly.

4. Fittingly, S. Chandrasekhar first discussed the picket-fence model in 1936 (*Monthly Notices of the Royal Astronomical Society* 96 [1936]: 21).

5. Or plasma physicists, or chemists, or biologists. Many scientists in many different fields use spectroscopy.

6. The spectra to be described fall under the generic label "thermal spectra" (spectra from a hot source). Once a label exists, the inverse label is usually stuck onto something: "nonthermal spectra" also exist. For now, we will close our eyes to this possibility.

7. For the advanced reader: the spectrum is named for the German physicist Max Planck (1858–1947). The Planck curve has a particular shape given by the equation $B_\nu(T) = (2h\nu^3/c^2)\,[e^{h\nu/kT} - 1]^{-1}$, where B = intensity of the radiation at the specified frequency, T = temperature, h = Planck's constant, ν = frequency, c = speed of light, and k = Boltzmann's constant. This equation tells us, for a specified temperature, the amount of radiation to be expected from a hot, opaque source, at every single wavelength (or equivalently, frequency) in the electromagnetic spectrum (even if the expected amount is minute).

8. Thought experiments are useful in physics. Basically, one tries to conjure up a description of a phenomenon and then reason one's way through its behavior in the context of the particular physics that controls the phenomenon. A thought experiment allows the user to strip out any extraneous or distracting effects. For the experiment under discussion in chapter 6, a real star or gas cloud is not a pure gas. We imagine it is for the sake of discussion. Perhaps the most famous thought experiment occurred when Albert Einstein tried to visualize what he would see if he traveled at the speed of light. His thoughts eventually led to the special theory of relativity.

9. The clever reader will see that gas may be present at some location in space but remain invisible to us because no continuum light lies behind it. Interstellar gas is difficult to detect for just this reason.

10. All the spectra discussed in the text also fall under the overall heading of emission spectra. This terminology can be confusing, but basically, in this context, an emission spectrum comes from a source that emits energy. The other basic type of spectrum is a reflection spectrum. To understand a reflection spectrum, return to our picket-fence model. Place a mirror so that it reflects the picket fence. The object (the mirror) itself does not emit a spectrum but reflects the light falling onto it. A perfect example sits in our sky two weeks of every month: the Moon. The optical spectrum of the Sun reflects off the Moon. The Moon contributes none of the light we see.

If we constructed a diagram of these terms, it might resemble the following:

Type of Emission	Emission Process
	continuum
Emission →	emission (lines)
	Absorption

Reflection

11. A considerable amount of stellar evolution has been ignored between this sentence and the previous one. Stars go through several stages of expansion and contraction as well as several pulsation phases as they exhaust their fuel. More information is available in, for example, J. Kaler, *Stars* (San Francisco: W. H. Freeman, 1998).

12. For the astrophysically literate reader, textbooks abound on this subject. The best are D. Mihalas, *Stellar Atmospheres* (San Francisco: W. H. Freeman, 1970), D. Emerson, *Interpreting Astronomical Spectra* (New York: J. Wiley and Sons, 1996), and F. Shu, *The Physics of Astrophysics* (Mill Valley, Calif.: University Science Books, 1991).

13. The technical term is "pressure broadening," which widens the line above the natural breadth of the line. The books cited just above describe pressure broadening. Pressure broadening results from atomic or electronic or ionic collisions, depending on the nature of the gas.

14. Light also behaves as a particle, depending on the nature of the experiment. This is the wave-particle duality that lies at the heart of quantum physics.

15. The optical spectrum shows SN1988H in NGC 5878 at an age of about 50 days after the explosion. Supernovae are labeled by the year the explosion was detected and are assigned letters in increments to indicate their place in the explosions detected that year. SN1988H was the eighth supernova detected in 1988. The advent of more supernova searches and searches to fainter optical magnitudes raised the average number of detected supernovae from about 25 per year in the early and mid-1980s to about 175 per year in the late 1990s.

16. The proper interpretation is uniform expansion of the entire atmosphere surrounding the optically thick core. The absorption line comes from the portion of the expanding atmosphere that lies between the core and us; the emission line comes from the surrounding annulus that is projected against the black of space.

17. From P. Fisher et al., *Astrophysical Journal* 143 (1966): 203. A note added in proof in the paper states that spectra obtained of Sco X–1 by other rocket teams demonstrate that no peak exists in the 2–12-keV band. A spectrum obtained with the European satellite EXOSAT does show a peak near 4 keV, suggesting that the Fisher et al. spectrum is correct; see White et al., *Astrophysical Journal* 296 (1985): 475.

18. For the electronically inclined reader: the CCD is constructed of a layer of conducting metal (p-type silicon), a depletion layer (n-type silicon), and a second layer of conducting metal. A potential difference is applied across the two conducting layers. The charge is transferred sequentially to the readout node. Selective doping of the silicon divides it into pixels, with heavily doped channel stops walling off one pixel from its neighbor. Additional information is available in M. Longair, *High Energy Astrophysics,* vol. 1, 2d ed. (Cambridge: Cambridge University Press, 1992), 251, or any good electronics book.

19. The iron line of MCG-6-30-15 is discussed in R. Blandford and N. Gehrels, "Revisiting the Black Hole," *Physics Today* 52, no. 6. An update that covers NGC 3516 is presented in *Physics Today* 52,10 (October 1999).

20. Equivalently, the data may be noisy because the effective collecting area of the telescope is too small. Increasing the effective collecting area of any telescope by significant factors increases the signal-to-noise ratio of the spectra (for the case of a photon-noise-limited observation). Signal-to-noise limits are discussed in Longair, *High Energy Astrophysics,* 241.

21. Compare the emission line from NGC 3516 with the emission line from the supernova (Fig. 6.8). Both have photon noise, but the supernova spectrum contains so many more photons that the variations of noise are about the size of the line that draws the spectrum. Recall that the energy carried by one X-ray photon approximately equals the energy of a thousand optical photons.

22. The gravitational redshift was first predicted by Albert Einstein based on his general theory of relativity. The total amount of the redshift is given by $z = (1/c^2)(GM/R)$, where z is the redshift, c is the speed of light, G is the constant of gravitation, M is the mass of the object, and R is the distance from the center of the mass to the gas cloud. The gravitational redshift of Earth is measurable, but small. The shift is larger for a white dwarf because the mass of the object is much larger (comparable to the mass of the Sun in a volume about the size of Earth).

23. The more slowly light moves in a medium, the more it bends at the boundary between the medium and the air. In other words, the frequency dependence of the speed of light in a medium affects the refraction of light in that medium.

24. C. Huygens developed a principle: each slit or opening acts essentially as a new source of light (or, for the ocean, of waves), leading to constructive and destructive interference. The basic equation is:

$$d \sin \theta = m \lambda$$

where d is the spacing of the slits or grating lines (or, for the ocean, channels or breakwaters), θ is the angle of observation as measured at the slit, m is the order, or number, of constructive interference peaks starting with zero as the peak directly in the path of the slit, and λ is the wavelength of the light.

25. The material on compact disks is from L. Bloomfield, *How Things Work* (New York: J. Wiley and Sons, 1997), 580.

26. B. McNamara et al., *Bulletin of the American Astronomical Society* 32 (2001): poster abstract 13.21.

27. Currently, mercury telluride produces the best results. For the interested reader: see C. Stahle, D. McCammon, and K. Irwin, "Quantum Calorimetry," *Physics Today* 52, no. 8 (August 1999): 32.

28. The basic equation governing the behavior of the calorimeter is $\Delta T = E_X/C$, where ΔT is the temperature change, E_X is the energy of the absorbed X ray, and C is the heat capacity of the absorbing material. The calorimeter should measure energies across the complete bandpass, say, for example, from 0.1 to 10 keV. To provide a large ΔT, C must be as small as possible.

29. Spectral lines have a natural line width that is always present because each energy level has a finite spread in energy resulting from Heisenberg's uncertainty principle. On top of the natural width, motion contributes additional width.

30. R. Antonucci and J. Miller, *Astrophysical Journal* 297 (1985): 621.

31. Astronomers postulate that active galaxies viewed along the polar axis are "blazars."

32. G. Madejski et al., *Astrophysical Journal* 535 (2000): 87.

33. The teams' results are described in Gebhardt et al., *Astrophysical Journal* 539 (2000): L13, and Ferrarese and Merritt, *Astrophysical Journal* 539 (2000): L9.

34. W. Israel, a black-hole researcher, describes the NGC 4258 result as "clinching evidence" and states that "if this [central object] is not a black hole, it would have to be something even more exotic." (W. Israel, *Publications of the Astronomical Society of the Pacific,* 112 [2000]: 583.)

Chapter 7

1. Conservation of energy may be the most fundamental behavior of our universe. Energy may be converted to matter and matter to energy ($E = mc^2$), but it is always conserved.

2. And you thought this was difficult!

3. The two worlds do unite in a sense, because the equations that describe quantum physics can be used to describe problems in the Newtonian world. Physicists of the 1920s recognized the apparent disparity and noted that at the boundary between the classical and quantum worlds, the two formulations had to agree. This is known as the correspondence principle. When the fine details of microscopic physics do not matter, the equations of quantum physics describe, or correspond to, the behavior of macroscopic physics.

4. The constant unit is called the riser; as anyone who has encountered stairs with risers of differing heights is aware, they are dangerous. The human brain becomes used to stairs of approximately identical heights. Shallow stairs, such as often are encountered at government buildings, for example, turn out to be agonizingly slow with their small risers and wide treads.

5. The number of protons essentially dictates the nuclear chemistry of an atom. The number of electrons largely dictates the electrical chemistry of an atom. The number of neutrons equals or is greater than the number of protons; atoms with the same atomic number but different atomic weights are known as isotopes. Both $_{92}U^{235}$ and $_{92}U^{238}$ are isotopes of uranium; because the element is identified as uranium by its chemical symbol

(U), the "92" that signifies the number of protons is essentially redundant information and usually not written.

6. Atomic-mass units are based on the mass of the carbon atom, which has six protons and six neutrons. This mass is defined to be exactly 12. The mass of a proton is 1.672×10^{-24} grams; for the neutron, the mass is 1.675×10^{-24} grams.

7. The arrangement of the periodic table highlights the similar behaviors among the chemical elements as shown by the Russian chemist D. Mendeleev around the middle of the 1800s.

8. The alert reader may ask the degree of distinctiveness of the set of emission or absorption lines. For example, the energy levels of an ionized atom of oxygen, now with seven electrons instead of the neutral eight, should lie in the same locations as the energy levels of the neutral nitrogen atom with its seven electrons. This does not occur, because the nucleus of the oxygen atom has more mass by the ratio 16/14. The extra mass alters the positions of the energy levels. There are situations, however, in which the emission or absorption lines of different elements do lie near one another. This occurs particularly for elements with many electrons (for example, iron), which have thousands of possible transitions between energy levels.

9. There are exceptions to this statement that would overly complicate the picture. Basically, each energy level, except the lowest, usually has more than one possible transition to a lower energy level. One of the transitions is the dominant one for a given range of temperatures or densities, dominant in the sense that the average time an electron remains in the higher level is short. However, there are conditions, usually of very low density, in which the dominant transition does not occur efficiently and a less efficient transition is the path to a lower energy level. The lifetime of an electron in the higher energy level is then considerably longer.

10. Prior to 1950 or so, the evolution of a given star was thought to move it either upward along the main sequence or downward. That these models did not accord with the facts was realized only with increasing observations in the 1940s. A star moves *away* from the main sequence as it runs short of fuel. (O. Gingerich, "The Summer of 1953: A Watershed for Astrophysics," *Physics Today* 47, no. 12 [December 1994], 34.)

11. Models of the interior structure of stars of various mass show that the proportion of hydrogen goes to zero somewhere between about a tenth of the mass of the star and about half. Astrophysicists who study the internal structure of a star do so by building computer models using the equations of stellar structure. These equations, reproduced below, dictate how the temperature, pressure, and energy generation change with distance from the center of the star. One can construct a model of a star that uses distance from the core in units of length (kilometers, for example). However, that is not the most meaningful unit; it is easier to work in the coordinate of mass that follows the fraction of mass from center to outer surface. One reason for this approach: comparing high-mass and low-mass stars is considerably easier because a particular behavior may occur at about the same location in mass, but not in physical distance from the core. For example, at the time a star of one solar mass has lived 85 to 90 percent of its total life, its innermost 10 percent, by mass, contains about 11 percent hydrogen, 86 percent helium, and 3 percent heavier elements. For a five-solar-mass star at the same point in its life, the innermost 10 percent of the star by mass contains essentially zero hydrogen. The point in the star's mass coordinate at which 11 percent of the matter is hydrogen occurs at a mass coordinate of about 58 percent. The hotter star burns more of the hydrogen farther outward than the low-mass star. Values for stellar models were taken from E. Novotny, *Introduction to Stellar Atmospheres and Interiors* (New York: Oxford University Press, 1973).

The equations of stellar structure:

Conservation of mass: $dM(r) / dr = 4\pi r^2 \rho(r)$;
Conservation of energy: $dL(r)/dr = 4\pi r^2 \rho(r)[\varepsilon(r) - T(r) \, dS/dt]$;
Conservation of momentum: $dP(r)/dr = -\rho(r) \, G \, M(r) / r^2$;

Transport of energy: $dI(r)/dr = -(3/4ac)\kappa(r)\rho(r)[L(r) / 4\pi r^2] T^{-3}(r)$ (radiative);
$$= (\Gamma_2 - 1)/\Gamma_2 \, (T(r) / P(r)) \, dP(r)/dr \text{ (convection)}$$

where the variables are: M = mass, r = radius, ρ = density, L = luminosity, ε = rate of energy generation, T = temperature, S = entropy, t = time, P = pressure, G = gravitational constant, κ = mean opacity of the matter, and Γ_2 = adiabatic exponent.

These equations must be supplemented by "microphysics" equations that describe individual nuclear reactions, atomic structure, and the like.

12. The layers of the star would not necessarily be so distinct as an onion's layers, but the analogy is appropriate.

13. The details of how the core collapse detonates the star are still debated. One of the current ideas is the creation of a huge number of neutrinos. The densities at the core of an exploding star are so high that many of the neutrinos transfer their energy and momentum into the layers just above the core, propelling it outward. More-recent work (for example, the research of J. Craig Wheeler and his colleagues at the University of Texas in Austin) suggests the creation of a jet that essentially rips a hole through the star. The debate continues.

14. The supernova was SN 1993J in NGC 3031 (= M81). The detection was reported by M. Leising et al., *Astrophysical Journal (Letters)* 431 (0000): L95. The OSSE detector (Oriented Scintillation Spectrometer Experiment) on the Compton Gamma-ray Observatory covered 50 keV to 10 MeV. The model spectrum required a high temperature (82^{+61}_{-29} keV) and density to be consistent with the observations.

15. SN1987A handed theorists a surprise by "being blue when it blew." (The phrase comes from a talk by R. Kirshner of Harvard University.) When stellar-evolution researchers construct interior models of stars, they also produce a description of how the model star appears to an observer. This description arrives as a set of temperatures and luminosities, both of which change with time. Connecting the dots formed by each pair of temperature and luminosity values creates a path or track in the Hertzsprung-Russell diagram. The track describes how the surface temperature and luminosity change with time as the star uses its fuel. Stellar tracks generally move from the main sequence toward the upper right of the diagram. Many of the stellar-evolution models thus predicted supernovae from red-giant stars. SN1987A showed that some blue stars might explode as supernovae. Additional mixing within the progenitor brought at least one stellar track back toward the blue.

16. A brief summary of the acoustic heating of the solar chromosphere is provided in H. Zirin, *Astrophysics of the Sun* (Cambridge: Cambridge University Press, 1988).

17. A brief look at the past 25 years of studies of the chromospheres and corona of the Sun and stars is contained in J. Drake's contribution to *X-ray Astronomy 2000: Proceedings of the Palermo Conference on X-ray Astronomy* (San Francisco: Astronomical Society of the Pacific, 2001). The summary of the acoustic theory's failings comes from B. Haisch, "The Solar-Stellar Connection," in *The Many Faces of the Sun*, ed. Keith Strong et al. (New York: Springer-Verlag, 1999), 484.

18. Observations of the Sun have become extremely detailed. Observations from TRACE, the Transition Region and Coronal Explorer satellite, SOHO, the Solar and Heliospheric Observatory, and the soft-X-ray telescope on the Japanese solar satellite Yohkoh have essentially revolutionized our view of the Sun. The wealth of detail in the images is nothing short of astounding. Details, including video loops, are available at <http://vestige.lmsal.com/TRACE/> (TRACE, Lockheed Martin), <http://www.lmsal.com/SXT/> (Yohkoh), and <http://sohowww.nascom.nasa.gov/> (SOHO).

Chapter 8

1. A typical X-ray photon has an energy of 1 keV = 1,000 eV; a typical optical photon has an energy of 1 eV.

2. The Planck curve peaks at about 4,800 to 5,000 Angstroms, in the blue/blue-green portion of the spectrum. The Planck curve falls relatively quickly to shorter wavelengths, but more slowly to longer wavelengths. This difference shifts the mean color slightly toward longer wavelengths, resulting in an overall yellow color. Furthermore, an area effect intrudes: at the center of the apparent disk, we see slightly more deeply into the Sun than at the edge; the deeper layers are slightly hotter, while the shallower layers are slightly cooler. The apparent area of the cooler regions seen by an observer on Earth is larger than the apparent area of the hotter regions.

3. Compton scattering has been used in medical diagnosis, particularly for osteoporosis: the intensity of scattered gamma rays can be directly related to the density of the scattering material. If osteoporosis afflicts the patient's bones, the intensity of scattered gamma rays drops.

4. Waves become shocks when their speed exceeds the speed of sound in the medium through which they travel. A sonic boom records the passage of a shock wave. Sound waves travel in air at about 1,193 kilometers per second (about 741 miles per hour). Jets exist that move more quickly in air than that speed. When a jet breaks the sound barrier, it moves through the air faster than the speed at which sound waves move. The pressure waves pile up in the direction of motion. The shock's arrival at ground level is a sonic boom.

5. I. Tuohy and M. Dopita describe the planar nature of the supernova remnant in *Astrophysical Journal* 268 (1983): L11.

6. Take two jars with openings of identical size. Fill one jar with salt water and the other with clear water. Place a card over the second jar, invert it, and place it on top of the salt-water jar. Slowly remove the card. The fresh water will sit on top of the salt water. Now, do the same experiment, but place the salt-water jar on top. Once the card is removed, "fingers" of salt water will develop until the salt water sits in the bottom jar.

7. The reader may argue that a relatively highly ionized ion will attract electrons and return to its more natural state of electrical neutrality. Said reader would be correct for a dense environment. The interstellar medium, however, is tenuous; ions may last for years in an excited state, whereas, in a dense environment, they would be stimulated and undergo a downward transition. I recall reading a vivid description of the low density of the interstellar medium: on a cold day, exhale; the amount of visible vapor, spread over a cube a kilometer on each side, yields a density about a thousand to a million times greater than the density of the interstellar medium.

8. For advanced readers: those lines are as follows for the simplest situation possible. Assume a spherical accretion flow at rate A onto a compact star of mass M and radius R. The gas will flow inward at the free-fall velocity until it hits the surface of the neutron star, at which point the gas will immediately decelerate. The infall velocity converts to heat as the energy and momentum of the infalling gas transfer to the neutron star's surface layers. The luminosity $L = GMA/R = \frac{1}{2} Mv^2$. But $L = 4\pi R^2 \sigma T^4$. For typical values of A, M, R for a white dwarf (1 solar mass, A about 10^{-9-11} solar masses per year, R about 5,000 kilometers), the temperature is about 3 eV, in the ultraviolet; for values typical of a neutron star (M = 1–2 solar masses, R = 10 kilometers, A about 10^{-9-11} solar masses per year), the temperature is 10^7 K, or about 1 keV.

9. The radiation was first detected in a synchrotron, one type of particle accelerator.

10. Within sunspots, the magnetic field is about a thousand gauss (see H. Zirin, *Astrophysics of the Sun* (Cambridge: Cambridge University Press, 1988).

11. A thousand trillion is 10^{15} with the prefix "peta." Other, more familiar prefixes include "mega" for 10^6, "giga" for 10^9, and "tera" for 10^{12}. The current target of high-speed computing is a peta-flop machine, one capable of 10^{15} calculations per second. "Trillion" is more familiar to us because of the budget of the United States Government. Beyond "peta" is "exa" for 10^{18}.

12. Laboratory magnetic-field strengths obtained from the Web site of the National Magnetic Field Laboratory at Florida State University, <www.magnet.fsu.edu>.

13. In astrophysics, one detected object is "unique," two produce a "puzzle," and three or more constitute a "class" of behavior.

14. The SGR discussed is known as SGR 1900+14. (See C. Kouveliotou et al., *Astrophysical Journal (Letters)* 510 (1999): L115.

15. D. Meier, S. Koide, and U. Yutaka, *Science* 291 (2001): 84.

16. The polarization detection was obtained by S. Tapia in 1976; see S. Tapia, *Astrophysical Journal (Letters)* 212 (1977): L125.

17. A recent study suggests that colliding neutron stars contribute a fraction of the chemical elements. See note 24 of chapter 4 (p. 176).

18. Because scientists like to create acronyms when needed, the proper name would be something like "SHGSs," for "Speckled Hot Gas Spheroids."

19. The Chandra observation of the Perseus cluster of galaxies perhaps shows the gaps most clearly. (See A. Fabian et al., *Monthly Notices of the Royal Astronomical Society* 318 (2000): L65.

20. We expect to have available someday soon two additional windows on the universe: routine observations from sensitive neutrino "telescopes" and gravity-wave observatories. Neutrinos have been detected from the Sun, albeit in lower numbers than expected, and from supernova 1987A in the Large Magellanic Cloud. Those detections merely whet the appetite. Gravity waves have not yet been detected, but their presence has been inferred from the changes in the orbit of the binary pulsar 1913+16. The changes precisely match the predictions of Einstein's general theory of relativity. The observations of the changes in the orbit of the binary pulsar won J. Taylor and R. Hulse the Nobel Prize in physics.

Chapter 9

1. Luckily, as an astrophysicist, I'm somewhat insulated from the frustration. People who ask me questions and do not receive straight answers seem to accept that scientists do not understand everything about the universe.

2. The field continues to be revitalized: recently a plastic has been discovered to exhibit properties of superconductivity (*New York Times,* March 8, 2001), and the fullerene compounds may also be superconductive.

3. Critics of weather forecasters, defined as anyone who does not earn a living as a weather forecaster, argue that, even with all the research, weather forecasters do not predict well. That may appear to be true, but weather forecasters suffer an occupational hazard, one that is a fundamental problem of human psychology: people remember only the days when the forecasters get it wrong. In science, this is known as a "selection effect."

4. "Semirandomly" because the reviewers must be knowledgeable about the subject of the proposal. Within that limitation, however, essentially anyone who has carried out research in the field and who is available during the review period is a potential reviewer.

5. The formal titles of the reviews are: *Ground-Based Astronomy: A Ten-Year Program,* National Research Council (Washington, D.C.: National Academy of Sciences, 1964) (the "Whitford report"); *Astronomy and Astrophysics for the 1970s,* National Research Council (Washington, D.C.: National Academy of Sciences, 1972) (the "Greenstein report"); *Astronomy and Astrophysics for the 1980s: Report of the Astronomy Survey Committee* (Washington, D.C.: National Academy Press, 1982) (the "Field' report"); *The Decade of Discovery in Astronomy and Astrophysics,* Astronomy and Astrophysics Survey Committee (Washington, D.C.: National Academy Press, 1991) (the "Bahcall report"); *Astronomy and Astrophysics in the New Millennium,* Astronomy and Astrophysics Survey Committee (Washington, D.C.: National Academy Press, 2001) (the "McKee/Taylor report").

6. The Bahcall report of 1991 actually had as the number one recommendation that funds be sought simply to support researchers, their students, and basic infrastructure. Research grants had become, by the late 1980s, so difficult to obtain that the lack of money for researchers to publish papers, to travel to science conferences, or to travel to observatories was itself hindering progress in the field.

7. For amateur telescopes, an oft-cited rule of thumb is one-eighth of the wavelength of optical light. For X rays of 1 to 10 angstroms (about 1 to 10 keV), that requirement

means polishing to about 0.1 angstroms. This precision, and the corresponding attention to detail that it requires, raise the cost of producing the mirror. See N. Howard, *Standard Handbook for Telescope Making* (New York: Thomas Crowell, 1959).

8. C. R. O'Dell, *Telescopes for the 1980s* (Palo Alto: Annual Reviews, 1981), 129.

9. Reported in M. Waldrop, *Science* 227 (1985): 285.

10. In contrast to a facetious comment I made earlier, the word "facility" was used intentionally. Funding to build Chandra sat before Congress at the same time that the mirror flaw of the Hubble Space Telescope was discovered. The word "observatory" was not in good graces at the time. (Private conversation with the Chandra Observatory director, Dr. Harvey Tananbaum.)

11. Among engineers, this is part of Wexelblatt's scheduling algorithm. The full text is: "Choose any two: Fast Cheap Good."

12. Critics of the space program often argue that we should not spend money "out there." They apparently do not notice that Congress funds space projects, which then hire people to work here on Earth; the workers spend their salaries "down here." The critics apparently mean we should not pursue investigations into space while we continue experiencing problems here on Earth. That same reasoning, applied at numerous historical junctions, would have stifled innovations that ultimately led to lower costs for all.

We have not yet reached the point where money is taken from the taxpayers and spent "out there." That time is coming, if we as a nation or a world decide to spend money by subsidizing the development of a separate economy at, for example, a permanent base on the Moon.

13. This is not so silly as it sounds: the Hubble Space Telescope was initially conceived by Dr. Lyman Spitzer Jr., a longtime faculty member at Princeton University, in the late 1940s. The telescope became a firm proposal in the late 1960s and was launched in 1990. Dr. Spitzer died in 1997 at the age of 82. He exceeded the average life span for a man born in 1914 by about 25 years; had he not done so, he would never have seen the fruition of his concept for a space telescope.

14. Notice I have not said much about "better." If Wexelblatt's scheduling algorithm (see note 11) truly describes projects, then choosing "cheap" and "fast" may force the taxpaying public and Congress to accept some project failures. The failures of the Mars Climate Observer and Mars Polar Lander were caused by inexperienced teams. Essentially, then, "cheap" and "fast" were chosen by default over "better." The lessons learned from these failures will place more emphasis on "better." See M. Dornheim, *Aviation Week and Space Technology*, April 3, 2000.

15. SIRTF will be placed into a Sun-centered orbit to drift slowly behind Earth. To work in the infrared band, the detectors must be very cold. The environment immediately surrounding Earth is rather warm because Earth gives off considerable heat, particularly in the infrared band. In a solar orbit, however, the cooling requirements are lower, so the satellite does not require so large a reservoir of helium. A smaller reservoir not only is cheaper, liquid helium being an expensive commodity, but requires a smaller rocket to launch the satellite, so launch costs drop.

16. Mars Pathfinder (1997), Mars Global Surveyor (still working), Mars Climate Orbiter (lost in 1999), Mars Polar Lander (lost in 1999), Mars Odyssey (launched April 9, 2001), two geology rovers (c. 2003), a possible sample return mission (not yet scheduled). See NASA Office of Space Science, *The Space Science Enterprise: Strategic Plan*, November 2000.

17. By comparison, in the next few years there will be so many missions that the Deep Space Network, used to communicate with all satellites, will be unable to handle the volume of commands. NASA officials are studying ways to alleviate this traffic jam. (A. Bridges at www.space.com, December 4, 2000.)

18. Is there a down-to-Earth equivalent? Absolutely. How much of an uproar would there be from fans if a sports team owner told them that they'd just have to wait 10 to 20 years for a winning team? (Note: that question does not apply to fans of the Boston Red Sox.)

19. On a superficial level, one could look at Venus, Mars, and the outer planets and state "No life there, life on Earth, question answered." However, the meteorology of Venus is

quite different from the meteorology of Earth, and both differ from the meteorology of Mars. Mars uniquely has a condensation flow from one pole region to another as the carbon dioxide melts at the summer pole and recondenses at the winter pole. The Hadley circulation, the large-scale circulation of air that, warmed by the Sun, rises from the equator and flows toward the poles, differs on the three terrestrial planets: Mars and Venus both have a single Hadley cell, while Earth has three cells; the multiplicity leads to the trade winds. The Great Red Spot of Jupiter has been compared to a long-lived hurricane. Studies of the Voyager images show that the spot lives between zonal jets moving in opposite directions. For additional details, refer to the atmospheric articles by A. Ingersoll and B. Jakosky in *The New Solar System*, 4th ed., ed. J. K. Beatty, C. C. Petersen, and A. Chaikin (Cambridge, Mass.: Sky Publishing, 1999).

20. Themes, questions, and road maps are all available at <www.oss.hq.nasa.gov>.

21. Yes, actual tape recorders (space qualified). An interesting problem with tape recorders: because they spin, that spin must be counted correctly (in a manner not so different from the one accountants use for money) or the entire satellite can acquire some of the spin. The relevant principle of physics is called the conservation of angular momentum; it is of critical importance in managing a satellite. Momentum must be "dumped" on a regular schedule; this dump time forms part of the overhead in satellite management.

22. For example, the data may be sent to NASA's Tracking and Data Relay Satellite system.

23. NASA received the first proposal for a satellite that became RXTE in the late 1970s. At that time, the most advanced chips available were 286s. At the time of RXTEs launched in late 1995, the best chips available were 486s and 586s, but RXTE still carried space-qualified 286s.

24. Chandra has an effective area of about 700 square centimeters at 1 keV; Newton's area is about 1500 square centimeters at 1 keV. Chandra's point-spread function delivers more than 70 percent of the X rays within 0.5 seconds of arc; the mirrors on Newton place about 50 percent within ~6 seconds of arc.

25. Perhaps the closest competitor is paleontology, particularly the study of dinosaurs.

Chapter 10

1. I leave the choice of the answerable question as an exercise to the reader.

2. See chapter 3, note 7, for some historical background.

3. The MMT was originally called the Multiple Mirror Telescope, as it was the first optical astronomy telescope to combine the light collected by smaller mirrors instead of using one large mirror. That experience paved the way for telescopes with mirrors such as the segmented ones used by Kecks I and II. Now the MMT has a single, spun-cast, 6.5-meter mirror. It retains the initials, so the "Multi" has been changed to "Mono."

4. This is the Ball Aerospace baseline configuration; see *Science with the Constellation-X Observatory*, NASA Brochure NP-1998-067-GSFC, January 1999.

5. This is the TRW baseline configuration; see NASA Brochure NP-1998-067-GSFC.

6. This is the GSFC-SAO baseline configuration; see NASA Brochure NP-1998-067-GSFC.

7. An exception to these comments is the Iridium constellation. Those satellites were in low-Earth orbit; consequently, the entire constellation was always in motion. The technical problem to be solved in designing Iridium was the rapid switching of a "phone call" from one satellite to another because coverage of any point on Earth was a property of the entire constellation and not of a particular component.

8. Recall that the gratings, discussed in chapters 6 and 7, diffract the light, reducing the signal-to-noise ratio.

9. Above 50 keV, the Oriented Scintillation Spectrometer Experiment on the Compton Gamma-ray Telescope provided very hard X-ray imaging. The field of view of OSSE was about 3.8 degrees by 11.4 degrees.

10. The list comes from B. Wilkes, "Quasars and Active Galactic Nuclei," in *Allen's Astrophysical Quantities,* 4th ed., ed. A. N. Cox (New York: Springer-Verlag, 2000), 587.

11. When I first learned about XEUS, I could not find anyone who knew what the abbreviation represented. I knew that if the proposal were chosen to become a real mission, words would be assigned to the letters. I assumed the letters stood for something like "X-rays from a European Union Satellite." XEUS actually stands for X-ray Evolving Universe Satellite, which defines the mission clearly.

12. The foil mirrors are described in P. Serlemitsos, *Applied Optics* 27 (1988): 1447. They are essentially thick aluminum foil overcoated with a thin layer of an X-ray-reflecting material such as gold. (The Chandra mirrors are overcoated with iridium.)

13. This explains why all X-ray telescopes resemble long, narrow tubes.

14. Formation flying is also required by gravity-wave researchers. The proposed LISA mission (Laser Interferometer Space Antenna) uses three spacecraft flying in a triangle. Lasers will link the three. Each spacecraft contains an independent, isolated proof mass. Optics and feedback circuits center the shell of each satellite on the isolated proof mass. As gravity waves propagate through the triangle, the three spacecraft will alter their relative positions slightly; the wave's passage will be detected by the feedback as each satellite re-centers itself around the proof mass. Gravity-wave observatories on the ground cover frequencies above a few hertz; LISA will extend the range to about 10^{-4} hertz. Where the ground observatories expect to see gravity signatures of core-collapse supernovae and collisions or coalescence of neutron stars, LISA extends our view to ordinary galactic binary systems. On NASA's Structure and Evolution of the Universe Roadmap Strategic Plan for 2003 to 2023, LISA falls into the 2003–7 time frame, with a probable launch near the end of that period. The ground observatories, called LIGO for Laser Interferometer Gravitational-Wave Observatory, should start collecting data "any day now." LIGO is described in Barish and Weiss (1999).

15. This section assumes that Con-X and XEUS will be built and flown. Con-X is the choice of the United States' X-ray astronomy community to follow Chandra, as is XEUS for the European community. Vagaries of funding and the economies of the United States and Europe may or may not allow these plans to unfold.

16. Perhaps it is just as well that we do not have eyes of micro-arc-second resolution. Can you imagine turning to your significant other and staring down into the pores of the skin? Our heads would probably explode with all the details our eyes would deliver to our brains.

17. Interferometry using X rays has already been demonstrated in the laboratory by Dr. Webster Cash and his coworkers of the University of Colorado (Cash et al. 2000).

18. $\theta = 251643 \lambda / d$, where θ = angular resolution in seconds of arc, λ = wavelength of light, and d = diameter of telescope aperture, with λ and d in identical units.

Bibliography

Abell, G. *Exploration of the Universe*. 4th ed. Philadelphia: Saunders College Publishing, 1982.

Abrash, M. "Color-modeling in 256-Color Mode." *Dr. Dobb's Journal* 17, no. 8 (August 1992): 149.

Adams, F., and G. Laughlin. "A Dying Universe: The Long-term and Evolution of Astrophysical Objects." *Reviews of Modern Physics* 69 (1997): 337.

Amelio, G. "Charge-coupled Devices." *Scientific American,* February 1974, 23.

Anderson, M., et al. "Relativistic Electron Populations in Cassiopeia A." *Astrophysical Journal* 373 (1991): 146.

Antonucci, R., and J. Miller. "Spectropolarimetry and the Nature of NGC 1068." *Astrophysical Journal* 297 (1985): 621.

Bahcall, J. *The Decade of Discovery in Astronomy and Astrophysics: Report of the Astronomy and Astrophysics Survey Committee*. Washington: National Academy Press, 1991.

Bahcall, J. N., and R. Davis, Jr. "An Account of the Development of the Solar Neutrino Problem." In *Essays in Nuclear Astrophysics*, ed. C. Barnes, D. Clayton, and D. Schramm, 243. Cambridge: Cambridge University Press, 1982. Reprinted in J. N. Bahcall, *Neutrino Astrophysics* (Cambridge: Cambridge University Press, 1989).

Bailyn, C., et al. "Dynamical Evidence for a Black-Hole in the Eclipsing X-ray Nova GRO:J1655-40." *Nature* 378 (1995): 157.

Barish, B. C., and R. Weiss. "LIGO and the Detection of Gravitational Waves." *Physics Today* 52, no. 10 (October 1999): 44.

Becker, R., et al. "Evidence for Reionization at $z\sim6$: Detection of a Gunn-Peterson Trough in a $z=6.28$ Quasar." *Astronomical Journal* 122 (2000): 2850.

Binney, J., and M. Merrifield. *Galactic Astronomy* (Princeton, N.J.: Princeton University Press, 1998).

Blandford, R., and N. Gehrels. "Revisiting the Black Hole." *Physics Today* 52, no. 6 (June 1999): 40.

Blanton, E., C. Sarazin, and J. Irwin. "Diffuse Gas and Low-Mass X-ray Binaries in the Chandra Observation of the S0 Galaxy NGC 1553." *Astrophysical Journal* 552 (2001): 106.

Bloomfield, L. *How Things Work*. New York: John Wiley and Sons, 1997.

Bohm-Vitense, E. *Introduction to Stellar Astrophysics*. Vol. 3, 86–100. New York: Cambridge University Press, 1992.

Bradt, H., T. Ohashi, and K. Pounds. "X-ray Astronomy Missions." *Annual Reviews of Astronomy and Astrophysics* 30 (1992): 391–427.

Carr, B., and J. Primack. "Dark Matter—Searching for MACHOS." *Nature* 345 (1990): 478.

Cash, W., et al. "Laboratory Detection of X-ray Fringes with a Grazing-Incidence Interferometer." *Nature* 407 (2000): 160.

Chandrasekhar, S. "The Radiative Equilibrium of the Outer Layers of a Star, with Special Reference to the Blanketing Effect of the Reversing Layer." *Monthly Notices of the Royal Astronomical Society* 96 (1936): 21.

Chen, W., C. Shrader, and M. Livio. "The Properties of X-Ray and Optical Light Curves of X-ray Novae." *Astrophysical Journal* 491 (1997): 312.

Ciotti, L., et al. "Winds, Outflows, and Inflows in X-ray Elliptical Galaxies. I." *Astrophysical Journal* 376 (1991): 380.

Clark, D. H., and F. R. Stephenson. *Historical Supernovae* (New York: Pergamon Press, 1977).

Clarke, A. C. "Extra-terrestrial Relays." *Wireless World* (October 1945): 305.

Das Gupta, S., and G. Westfall. "Probing Dense Nuclear Matter in the Laboratory." *Physics Today* 46, no. 5 (May 1993): 34.

Day, C. "New Results Suggest X-ray Emission Is a Common Property of Comets." *Physics Today* 50, no. 12 (December 1997): 21.

Day, C. "Long X-ray Observation Probes Black Hole Infall." *Physics Today* 52, no. 10 (October 1999): 24.

Dennerl, K., et al. "The First Broad-band X-ray Images and Spectra of the 30 Doradus Region in the LMC." *Astronomy and Astrophysics* 365 (2001): L202.

Dornheim, M. "NASA Says MPL [MPL = Mars Polar Lander] Was Too Cheap, Too Fast." *Aviation Week and Space Technology* (April 3, 2000): 40.

Drake, J. "Stellar Coronae: The First 25 Years." In *X-ray Astronomy 2000*, ed. R. Giacconi, S. Serio, and L. Stella, 53. San Francisco: Astronomical Society of the Pacific, 2001.

Dreyer, J. L., and R. Sinnott. *The Complete New General Catalog and Index Catalogues of Nebulae and Star Clusters*. Cambridge, Mass.: Sky Publishing, 1988.

Emerson, D. *Interpreting Astronomical Spectra*. New York: J. Wiley and Sons, 1996.

Fabian, A., et al. "Chandra Imaging of the Complex X-ray Core of the Perseus Cluster." *Monthly Notices of the Royal Astronomical Society* 318 (2000): L65.

Ferrarese, L., and D. Merritt. "A Fundamental Relation between Supermassive Black Holes and Their Host Galaxies." *Astrophysical Journal* 539 (2000): L9.

Field, G. *Astronomy and Astrophysics for the 1980s: Report of the Astronomy and Astrophysics Survey Committee*. Washington: National Academy Press, 1982.

Fisher, P., et al. "Observations of Cosmic X-Rays." *Astrophysical Journal* 143 (1966), 203.

Frank, J., A. King, and D. Raine. *Accretion Power in Astrophysics*. 2nd ed. New York: Cambridge University Press, 1992.

Freyberg, M., and R. Egger. *Proceedings of the Symposium "Highlights in X-ray Astronomy,"* ed. B. Aschenbach and M. Freyberg, 278. Garching: Max-Planck-Institut, 1999.

Gaetz, T., et al. "Chandra X-Ray Observatory Arcsecond Imaging of the Young, Oxygen-rich Supernova Remnant 1E 0102.2-7219." *Astrophysical Journal* 534 (2000): L47.

Garmire, G., et al. "The LogN-LogS Relation for the HDF-N Region As Observed by the Chandra X-ray Observatory." *Bulletin of the American Astronomical Society* 32 (2000): abstract 261.

Gebhardt, K., et al. "A Relationship between Nuclear Black Hole Mass and Galaxy Velocity Dispersion." *Astrophysical Journal* 539 (2000): L13.

Giacconi, R., et al. "First Results from the X-Ray and Optical Survey of the Chandra Deep Field South." *Astrophysical Journal* 551 (2001): 624.

Gilliland, R. "Hydrodynamical Modeling of Mass Transfer from Cataclysmic Variable Secondaries." *Astrophysical Journal* 292 (1985): 522

Gingerich, O. "The Summer of 1953: A Watershed for Astrophysics." *Physics Today* 47, no. 12 (December 1994): 34.

Gursk, H., et al. "A Measurement of the Location of the X-Ray Source SCO X-1." *Astrophysical Journal* 146 (1966): 310.

Haisch, B. "The Solar-Stellar Connection." In *The Many Faces of the Sun*, ed. Keith Strong et al., 481. New York: Springer-Verlag, 1999.

Hartmann, R., et al. "The Third EGRET Catalog of High-Energy Gamma-Ray Sources." *Astrophysical Journal Supplement* 123 (1999): 79.

Harwit, M. *Cosmic Discovery*. Brighton, Eng.: Harvester Press, 1981.

Hasegawa, I. "Catalogue of Ancient and Naked-eye Comets." *Vistas in Astronomy* 24 (1980): 59–102.

Hasinger, G., et al. "XMM-Newton Observation of the Lockman Hole. I. The X-ray Data." *Astronomy and Astrophysics* 365 (2001): L45.

Hayashi, C. "Evolution of Protostars." *Annual Reviews of Astronomy and Astrophysics* 4 (1966): 171.

Henize, K. "Catalogues of H-alpha-Emission Stars and Nebulae in the Magellanic Clouds." *Astrophysical Journal Supplement* 2 (1956): 315.

Howard, N. *Standard Handbook for Telescope Making*. New York: Thomas Crowell, 1959.

Ingersoll, A. *The New Solar System*. 4th ed., ed. J. K. Beatty, C. C. Petersen, and A. Chaikin, 201. Cambridge, Mass.: Sky Publishing, 1999.

Israel, W. "Black Hole 2000: The Astrophysical Era." *Publications of the Astronomical Society of the Pacific* 112 (2000): 583.

Jackson, J. D. *Classical Electrodynamics*. 2d ed. New York: John Wiley and Sons, 1975.

Jakosky, B. *The New Solar System*. 4th ed., ed. J. K. Beatty, C. C. Petersen, and A. Chaikin, 175. Cambridge, Mass.: Sky Publishing, 1999.

Kaler, J. *Stars*. San Francisco: W. H. Freeman, 1998.

Kouveliotou, C., et al. "Discovery of a Magnetar Associated with the Soft Gamma Repeater SGR 1900+14." *Astrophysical Journal* 510 (1999): L115.

Kraft, R., et al. "A Chandra High-Resolution X-ray Image of Centaurus A." *Astrophysical Journal* 531 (2000): L9.

Krauss, J. *Radio Astronomy*. New York: McGraw-Hill, 1966.

Lanford, W. "Ionization Chambers." In *McGraw-Hill Encyclopedia of Physics*, ed. S. Parker, 492–93. New York: McGraw-Hill, 1983.

Leising. M., et al. "Hard X-rays from SN 1993J." *Astrophysical Journal (Letters)* 431 (1994): L95.

Lewin, W. "GRO 51744-28; The Rapid Burster; Bizarre Objects!" In *The Hot Universe*, ed. K. Koyama, S. Kitamoto, and M. Itoh, 111. Dordrecht: Kluwer Academic Publishing, 1998.

Lisse, C., et al. "Discovery of X-ray and Extreme Ultraviolet Emission from Comet C/Hyakutake 1996 B2." *Science* 274 (1996): 205.

Lisse, C., et al. "Charge Exchange-Induced X-Ray Emission from Comet C/1999 S4 (LINEAR)." *Science* 292 (2001): 13343.

Longair, M. *High Energy Astrophysics*, vol. 1, 2d ed. Cambridge: Cambridge University Press, 1992.

Madejski, G., et al. "Structure of the Circumnuclear Region of Seyfert 2 Galaxies Revealed by Rossi X-ray Timing Explorer Hard X-ray Observations of NGC 4945." *Astrophysical Journal* 535 (2000): 87.

Maíz-Apellániz, J. "The Origin of the Local Bubble." *Astrophysical Journal (Letters)* 560 (2001): L83.

Masetti, N., et al. "Hard X-rays from Type II Bursts of the Rapid Burster and Its Transition toward Quiescence." *Astronomy and Astrophysics* 363 (2000): 188.

McCammon, D., and W. Sanders. "The Soft X-ray Background and Its Origins." *Annual Review of Astronomy and Astrophysics* 28 (1990): 657.

McClintock, J. "Probing Strong Gravitational Fields in X-ray Novae." In *Accretion Processes in Astrophysical Systems*, ed. S. Holt and T. Kallman, 290. New York: American Institute of Physics, 1998.

McKee, C., and J. Taylor. *Astronomy and Astrophysics for the New Millenium: Report of the Astronomy and Astrophysics Survey Committee*. Washington: National Academy Press, 2001.

McNamara, B., et al. "Discovery of Ghost Cavities in the X-Ray Atmosphere of Abell 2597." *Astrophysical Journal* 562 (2001): 149.

Meier, D., S. Koide, and U. Yutaka. "Magnetohydrodynamic Production of Relativistic Jets." *Science* 291 (2001): 84.

Mihalas, D. *Stellar Atmospheres*. San Francisco: W. H. Freeman, 1970.

Mushotzky, R., et al. "Resolving the Extragalactic Hard X-ray Background." *Nature* 404 (2000): 459.

Nandra, K., et al. "The Properties of the Relativistic Iron K-Line in NGC 3516." *Astrophysical Journal* 523 (1999): L17.

Novotny, E. *Introduction to Stellar Atmospheres and Interiors*. New York: Oxford University Press, 1973.

O'Dell, C. R. "The Space Telescope." In *Telescopes for the 1980s*, 129. Palo Alto, Calif.: Annual Reviews, 1981.

Oppenheimer, B., et al. "Direct Detection of Galactic Halo Dark Matter." *Science* 292 (2001): 698.

Paczynski, B., J. Ziolkowski, and A. Zytkow. "On the Time-Scale of the Mass Transfer in Close Binaries." In *Mass Loss from Stars,* ed. M. Hack, 237. Dordrecht: Reidel, 1969.

Peacock, J. *Cosmological Physics.* Cambridge: Cambridge University Press, 1999.

Peebles, P. J. *Principles of Physical Cosmology.* Princeton, N.J.: Princeton University Press, 1993.

Pischel, G. *A World History of Art.* 2nd rev. ed. New York: Newsweek Books, 1978.

Prendergast, K., and E. Spiegel. "Photon Bubbles." *Comments on Astrophysics and Space Physics* 5, no. 2 (1973): 43.

Rabinowitz, D., et al. "A Reduced Estimate of the Number of Kilometre-sized Near-Earth Asteroids." *Nature* 403 (2000): 165.

Ramsey, B., R. Austin, and R. Decher. "Instrumentation for X-ray Astronomy." *Space Science Reviews* 69 (1994): 139–204.

Rees, M. "Appearance of Relativistically Expanding Radio Sources." *Nature* 211 (1966): 468.

Rieke, G. *Detection of Light from the Ultraviolet to the Submillimeter.* Cambridge: Cambridge University Press, 1994.

Rutledge, R., et al. "Chandra Detection of an X-Ray Flare from the Brown Dwarf LP 944-20." *Astrophysical Journal* 533 (2000): L141.

Sagan, C., and A. Druyan. *Comet.* New York: Pocket Books, 1985.

Sambruna, R., et al. "Correlated Intense X-Ray and TEV Activity of Markarian 501 in 1998 June." *Astrophysical Journal* 538 (2000): 127.

Sarazin, C., J. Irwin, and J. N. Bregman "Chandra X-ray Observations of the X-ray Faint Elliptical Galaxy NGC 4697." *Astrophysical Journal* 556 (2001): 533.

Schlegel, E., R. Petre, and M. Loewenstein. "ROSAT Observations of X-ray-faint S0 galaxies—NGC 1380." *Astronomical Journal* 115 (1998): 525.

Schwartz, D., et al. "Chandra Discovery of a 100 Kiloparsec X-Ray Jet in PKS 0637-75." *Astrophysical Journal* 540 (2000): 69.

Science with the Constellation-X Observatory. Brochure no. NP-1998-067-GSFC. Washington: NASA, 1999.

Seliger, H. "Wilhelm Conrad Roentgen and the Glimmer of Light." *Physics Today* 48, no. 11 (November 1995).

Serlemitsos, P. "Conical Foil X-ray Mirrors: Performance and Projections." *Applied Optics* 27 (1988): 1447.

Shu, F. *The Physics of Astrophysics.* Mill Valley, Calif.: University Science Books, 1991.

The Space Science Enterprise: Strategic Plan. Washington: NASA, 2000.

Stahle, C., D. McCammon, and K. Irwin. "Quantum Calorimetry." *Physics Today* 52, no. 8 (August 1999): 32.

Stephenson, F. R. *Applied Historical Astronomy,* Joint Discussion 6. Twenty-fourth Meeting of the International Astronomical Union, Manchester, England, August 2000.

Tanaka, Y., and N. Shibazaki. "X-ray Novae." *Annual Reviews of Astronomy and Astrophysics* 34 (1996): 639.

Tapia, S. "Discovery of a Magnetic Compact Star in the AM Herculis/3U 1809+50 System." *Astrophysical Journal (Letters)* 212 (1977): L125.

Tozzi, P., et al. "New Results from the X-ray and Optical Survey of the Chandra Deep Field-South: The 300 Kilosecond Exposure. II." *Astrophysical Journal* 562 (2001): 42.

Tucker, W., and R. Giacconi. *The X-ray Universe.* Cambridge, Mass.: Harvard University Press, 1985.

Tuma, J. *Handbook of Physical Calculations.* 2nd ed. New York: McGraw-Hill, 1983.

Tuohy, I., and M. Dopita. "Ring Ejection in Type II Supernovae - 1E 0102.2-7219 in the Small Magellanic Cloud." *Astrophysical Journal* 268 (1983): L11.

Turner, M., et al. "Xeus: an X-ray Observatory for the Post-XMM Era." In *Proceedings of the Next Generation of X-ray Observatories Workshop,* ed. M. Turner and M. Watson. Published online, 1997. <http://ledas-www.star.le.ac.uk/ngxo/>

Turner, M., et al. "XMM-Newton First-light Observations of the Hickson Galaxy Group 16." *Astronomy and Astrophysics* 365 (2001): L110.

van Paradijs, J. "A Catalogue of X-Ray Binaries." In *X-ray Binaries,* ed. W. Lewin, J. van Paradijs, and E. P. J. van den Heuvel, 536. Cambridge: Cambridge University Press, 1995.

Vikhlinin, A., M. Markevitch, and S. Murray. "A Moving Cold Front in the Intergalactic Medium of A3667." *Astrophysical Journal* 551 (2001): 160.

———. "Chandra Estimate of the Magnetic Field Strength Near the Cold Front in A3667." *Astrophysical Journal* 549 (2001): L47.

Waldrop, M. "Astronomy and the Realities of the Budget." *Science* 227 (1985): 285.

Walter, F., and L. Matthews. "The Optical Counterpart of the Isolated Neutron Star RX J185635-3754." *Nature* 389 (1997): 358.

White, N., A. Peacock, and B. G. Taylor. "EXOSAT Observations of Broad Iron K Line Emission from Scorpius X-1." *Astrophysical Journal* 296 (1985), 475.

Wilkes, B. "Quasars and Active Galactic Nuclei." In *Allen's Astrophysical Quantities,* 4th ed., ed. A. N. Cox, 587. New York: Springer-Verlag, 2000.

York, D., et al. "The Sloan Digital Sky Survey: Technical Summary." *Astronomical Journal* 120 (2000): 1579.

Zirin, H. *Astrophysics of the Sun.* Cambridge: Cambridge University Press, 1988.

Suggested Readings

To learn more about X-ray astrophysics, the following articles or books may be valuable. *Scientific American* frequently publishes articles on astrophysics, of which articles devoted to X-ray astrophysics form a small portion. Recent *Scientific American* articles on astrophysics, arranged by general topic, include:

Stars, Stellar Evolution, Binary Stars

Basri, G. "The Discovery of Brown Dwarfs" *Scientific American*, April 2000.

Cannizzo, J., and R. Kaitchuck. "Accretion Disks in Interacting Binary Stars." *Scientific American*, January 1992.

Chaboyer, B. "Rip Van Twinkle: Oldest Stars Growing Younger." *Scientific American*, May 2001.

Fishman, G., and D. Hartmann. "Gamma-ray Bursts." *Scientific American*, July 1997.

Heuvel, E. P. J. van den, and J. van Paradijs. "X-ray Binaries." *Scientific American*, November 1993.

Kahabka, P., E. P. J. van den Heuvel, and S. Rappaport. "Supersoft X-ray Stars and Supernovae." *Scientific American*, February 1999.

Lasota, J. P. "Unmasking Black Holes." *Scientific American*, May 1999.

Piran, T. "Binary Neutron Stars." *Scientific American*, May 1995.

Galaxies

Beck, S. "Dwarf Galaxies and Starbursts." *Scientific American,* June 2000.

Bothun, G. "The Ghostliest Galaxies." *Scientific American*, February 1997.

Macchetto, F. D., and M. Dickinson. "Galaxies in the Young Universe." *Scientific American*, May 1997.

Veilleux, S., G. Cecil, and J. Bland-Hawthorn. "Colossal Galactic Explosions." *Scientific American*, February 1996.

Clusters of Galaxies

Henry, J. P., U. Briel, and H. Bohringer. "The Evolution of Galaxy Clusters." *Scientific American*, December 1998.

P. Charles and F. Seward, *Exploring the X-ray Universe* (Cambridge: Cambridge University Press, 1995), was written just after the launches of ROSAT and ASCA but before Chandra. It is a virtual encyclopedia of material on X-ray astronomy by two eminent scientists. The book is at about the same level as *Scientific American* articles and may be used in introductory college astronomy courses.

Physics Today occasionally publishes articles on X-rays and astrophysics. This magazine is the publication of the American Physical Society. Recent articles include:

Bennett, C., M. Turner, and M. White. "The Cosmic Rosetta Stone." *Physics Today* 50, no. 11 (November 1997): 32.

Bildsten, L., and T. Strohmayer. "New Views of Neutron Stars." *Physics Today* 52, no. 2 (February 1999): 40.

Blandford, R. and N. Gehrels. "Revisiting the Black Hole." *Physics Today* 52, no. 6 (June 1999): 40.

Helfand, D. "X-rays from the Rest of the Universe." *Physics Today* 48, no. 11 (November 1995): 58.

A history of the Einstein program is offered in W. Tucker and R. Giacconi, *The X-ray Universe* (Cambridge, Mass: Harvard University Press, 1985).

Sky and Telescope, August 1999, contains a prelaunch summary of Chandra written by Dr. Martin Elvis of the Smithsonian Astrophysical Observatory.

For a history of the development and construction of Chandra, see Wallace Tucker and Karen Tucker, *Revealing the Universe: The Making of the Chandra X-ray Observatory* (Cambridge, Mass.: Harvard University Press, 2001).

The history of the satellites used to study the X-ray universe is given in an article written by H. Bradt, T. Ohashi, and K. Pounds, *Annual Reviews of Astronomy and Astrophysics* 30 (1992): 391. The three authors each helped get X-ray astronomy off the ground.

A book describing the results of the ROSAT mission is available: B. Aschenbach, J. Hahn, and J. Truemper, *The Invisible Sky: ROSAT and the Age of X-ray Astronomy* (New York: Copernicus, 1997). Two of the writers had an enormous impact on the success of that mission.

Finally, there are numerous World Wide Web sites available for perusal and study. Only a few are listed here; all of these sites contain links to a wider Web presence for astronomy and astrophysics.

Astronomy-Picture of the Day: <http://antwrp.gsfc.nasa.gov/apod>

Chandra Science Center (public outreach): <http://chandra.harvard.edu>

High Energy Astrophysics Science Archive Research Center (HEASARC), NASA-GSFC: <http://heasarc.gsfc.nasa.gov/docs/corp/outreach.html>

Rossi X-ray Timing Explorer All-Sky Monitor movie: <http://xte.mit.edu/XTE/movie>

XMM-Newton: <http://xmm.vilspa.esa.es>

Additional Figure Credits

References to the Digitized Sky Survey of the Space Telescope Science Institute include the following additional credits: the Digitized Sky Survey used photographic date obtained using the Oschin Schmidt Telescope on Palomar Mountain. The Palomar Observatory Sky Survey was funded by the National Geographic Society. The Oschin Schmidt Telescope is operated by the California Institute of Technology and the Palomar Observatory. The plates were processed into the present compressed digital format with their permission. The Digitized Sky Survey was produced at the Space Telescope Science Institute under U.S. government grant NAG W-2166.

Index

A star, 99–100
Abell 3667 (A3667), 139–40
ABRIXAS, 165 n. 1
absorption, 94–101, 114, 118–21, 127–28, 132, 166 n. 12, 172 n. 27, 181 n. 16, 183 n. 8
absorption spectrum, 99–100
acceleration, 82–83, 116–18, 130, 132–33
accretion disk, 60–61, 83–84, 110–11, 134–37, 179 n. 31
accretion torus, 157
acoustic heating, 127, 184 n. 16
active galaxies (AGN), 9–10, 25, 34–36, 43, 54–55, 65–67, 98–99, 104–5, 110–13, 135–36, 157–58, 166 n. 12, 170 n. 45; blazars, 43, 157, 182 n. 31; Seyfert, 157; quasars, 65–67, 144, 157, 174 n. 7
All-Sky Monitor (XTE), 88, 179 n. 41
altitude, 10–11, 19–21, 79–80, 167 n. 20, 171 n. 9, 177 n. 6
AM Her stars (cataclysmic variables), 136–37
American Association of Variable Star Observers, 86–87
American Science and Engineering, 25
angstroms, 40
angular momentum, 83, 133, 188 n. 21
angular resolution, 44–45, 189 n. 18
Antonucci, R., 110–11, 182 n. 30
aphelion, 70
apogee, 76–77, 177 n. 8
Apollo 11, 166 n. 3
Aquilae X-1 (Aql X-1) (X-ray binary), 87–88, 179 n. 34

Ariel V, 23–24, 168 n. 24
arrival time, 4–5, 12, 15, 76–77, 79–81, 129, 134–35
Advanced Satellite for Cosmology and Astrophysics (ASCA), 31, 56–58, 104–5, 113, 122–24, 155–59, 173 n. 34, 38
Ashby, Jeffrey, 4–5
ASCA. See Advanced Satellite for Cosmology and Astrophysics
ASM (All-Sky Monitor), 88, 179 n. 41
Astro-1/BBXRT shuttle mission, 158
Astro-E observatory, xiii–xiv, 3, 10–11, 31, 109–10, 124–26, 143, 152–53, 242, 165 n. 1, 166 n. 12, 14
Astronomical Netherlands Satellite (ANS), 168 n. 24
atmosphere, xiii–xiv, 4–5, 10–11, 17, 19–21, 41, 43–44, 52, 100, 127, 133–34, 167 n. 20, 171 n. 16, 179 n. 36, 39, 181 n. 16
atomic mass, 15, 17, 118–19, 160, 183 n.6
atoms, 7, 12–15, 17–18, 28–29, 37, 39–46, 56–57, 65, 72, 74, 76–77, 81–85, 87–88, 88–90, 98–100, 102–7, 114–20, 125–31, 132–37, 182 n. 5, 183 n. 8, 11
aurora, 88–89, 133–34, 137, 166 n. 9, 178 n. 28, 179 n. 39
AXAF (Advanced X-ray Astrophysics Facility) (Chandra), 67–68, 165 n. 1, 166 n. 5, 178 n. 21

B star, 87–88
Bahcall, John, 28–29, 146–47; Bahcall Report, 146–47, 186 n. 5–6

Ball Aerospace and Technologies Corporation, x–xi, 188 n. 4

Bardeen, J., 144

beam dump, 135–36

Bednorz, George, 144–45

Bell Labs (AT&T), 56–57

BeppoSAX (satellite), 89–90

Big Bang, 93, 176 n. 24

binary stars, 25–26, 36, 47–48, 80–88, 90–91, 102–4, 109–10, 112–13, 136–37, 177 n. 17; cataclysmic variables, 36, 84–85, 112–13, 156–57; X-ray binaries, 31, 48–49, 87–88, 88–91, 109–11, 159–61, 168 n. 29; X-ray nova, 85–88

black holes, 7, 10, 23–24, 33–35, 65–66, 74, 83–86, 88–89, 104–5, 110–13, 147, 157, 163, 170 n. 45, 174 n. 8, 178 n. 22–24, 182 n. 34

blazars, 43, 157, 182 n. 31

Bohr, Niels, 116–20

bremsstrahlung (braking radiation), 130

brightness, 12, 56–57, 61–67, 69–70, 72–73, 77–79, 84–85, 88–96, 116, 139–40, 174 n. 9

Broadband X-ray Telescope, 173 n. 38

brown dwarfs, 74–75

Calibration Facility (XRCF), 67–69, 175 n.12–14

Callisto, ix

calorimeter, 15, 108–10, 125–26, 151, 157–58, 182 n. 28

Cannon, Annie, 116–17, 119–20

Cash, Webster, 189 n. 17

Cassiopeia A (Cas A) (supernova remnant), 7–9, 12–14, 45–46, 122–25, 127–28, 138, 140, 142

cataclysmic variable, 36, 84–85, 112–13, 156–57

CCD (Charge-Coupled Devices), 5–7, 15, 56–58, 67, 101–10, 122–25, 133–34, 151, 155–56, 173 n. 31, 33, 35, 37–38, 174 n. 39, 181 n. 18

CDF. *See* Chandra Deep Field

Centaurus A (Cen A)(AGN), 135–36

Centaurus X-2 (Cen X-2) (X-ray binary), 21, 23

Centaurus X-3 (Cen X-3) (X-ray binary), 88–90, 179 n. 34

centrifugal acceleration, 82–83

Chandra Deep Field (North, South) (CDF), 54

Chandra observatory, x, xiii–xiv, 3–7, 8–9, 12–14, 31–34, 37–40, 44–55, 59–61, 65–70, 72–75, 83–84, 88–91, 101–4, 106–8, 112–13, 122–30, 133–34, 138–40, 143, 146–47, 149, 151–62, 165 n.1, 165 n. 3, 172 n. 17, 173 n. 28, 178 n. 21, 186 n. 19, 187 n. 10, 188 n. 24, 189 n. 12, 15

Chandrasekhar mass limit, 83–84

Chandrasekhar, S., 83–84, 242, 165 n. 3, 178 n. 21, 180 n. 4

chromosphere, 127, 184 n. 16–17

cluster (galaxies), 25–28, 47–48, 66, 80–81, 107–8, 120, 138–142, 159–62, 186 n. 19

CMOS (Complementary Metal Oxide Semiconductor), 56–57, 173 n. 32, 37, 174 n. 39

CNO tri-cycle, 168 n. 32

collimator, 18–19

Collins, Eileen, 4–5

color, 10, 37, 41–43, 105–6, 116, 122–25, 129–30, 166 n. 7, 171, n. 12, 185 n. 2

Columbia (space shuttle), 3–5, 177 n. 6

Comet LINEAR (1999/S4), 72–73

Comet Shoemaker-Levy 9, 72–73

comets, 69–73, 159–60; xrays, 165 n. 2

Complementary Metal Oxide Semiconductor (CMOS), 56–57, 173 n. 32, 37–38

compression, 130–31

Compton, Arthur H., 166 n. 7, 176 n. 26, 185 n. 3

Compton (Gamma-Ray) observatory, xiii–xiv, 41–43, 74, 130, 149, 166 n. 7, 178 n. 26, 184 n. 14, 188 n. 9; Oriented Scintillation Spectrometer Experiment (OSSE), 188 n. 9

Compton scattering, 74, 130, 176 n. 27, 185 n. 3

conservation of energy, 28–29, 182 n. 1, 183 n. 11

constellation: Coma, 25; Fornax, 74–75; Monoceros, 84–85; Orion, 59–60, 65–67; Scorpius, 19–21, 52, 172 n. 24; Virgo, 25, 111

Constellation-X (Con-X). *See* High Throughput X-ray Satellite

convection, 126–27, 178 n. 19, 183 n. 11
cooling flow, 107–8
Cooper, L., 144
Copernicus (satellite), 137–38
cosmic structure and evolution, 159
Crab Nebula, 9–10, 21–23, 26, 32–33, 90, 134
Cygnus X-1 (Cyg X-1) (X-ray binary), 23–24, 179 n. 32

dark matter, xiv, 28–30, 138–39
Deep Space Network, 187 n. 17
de-excitation, 115–16, 119–20
degenerate matter, 99–100
density, 14–15, 17, 30–31, 60–61, 74, 81, 86–90, 95–99, 110, 119–21, 124–26, 130–32, 134, 138–40, 144–45, 157, 177 n. 13, 183 n. 9, 11, 184 n. 14, 185 n. 3, 7
detector, xiii–xiv, 4–7, 12, 15–21, 23–25, 30, 33–34, 37–40, 43–47, 52, 57–58, 62–63, 67–71, 76–77, 88–91, 94–95, 102–7, 121–22, 124–26, 133–34, 147–49, 151–52, 155–56, 163, 167 n. 21, 168 n. 31, 169 n. 34–35, 173 n. 38, 174 n. 3–4, 175 n. 12, 184 n. 14, 187 n. 15
diffraction, 100–102, 105–6, 162
diffuse X-ray background, 19–21
distortion, 38–39, 58, 131
Doppler, Christian, 100
Doppler shift, 100–101, 104–5, 111–13, 118–19, 127–28, 174 n. 7
downward transition, 114–15, 185 n. 7
dust clouds, 61–62

E0102-72 (supernova remnant), 131–32
Eagle Nebula, xiv, 92–93
Earth, x, xiii–xiv, 4–5, 7–11, 17, 34–35, 41–43, 52, 61, 70, 74–75, 78–85, 88–89, 100, 104–5, 114, 133–34, 137–38, 150–51, 156, 170 n. 48, 171 n. 15–16, 177 n. 5–6, 178 n. 18, 179 n. 35–36, 39, 181 n. 22, 185 n. 2, 187 n. 12, 15, 19, 188 n. 7
Eastman Kodak Company, x–xi
Eddington, Sir Arthur, 144
EGRET instrument (GRO), 41–43, 171 n. 13
Einstein, Albert, 30–31, 65–66, 115, 144, 180 n. 8, 181 n. 22, 186 n. 20

Einstein Observatory, x, 12–14, 22–23, 30–32, 38–39, 45–48, 49–51, 102–4, 126–28, 152–53, 160, 169 n. 40, 172 n. 17
electric current, 17–18
electric field, 17–19, 136–37
electromagnetic radiation, 40–43, 71–72
electromagnetic spectrum, 14–15, 41–43, 74, 76–77, 94–95, 116, 134–35, 149–50, 165 n. 2, 168 n. 23, 180 n. 7
electromagnetic waves, xiii, 41–43
emission, 16, 21–28, 31, 43, 47–54, 66, 70–74, 77, 81–82, 90–91, 97–105, 111–25, 127–42, 155–57, 159–63, 166 n. 12, 174 n. 9, 176 n. 21, 25, 179 n. 36, 180 n. 10, 181 n. 16, 21, 183 n. 8;
emission line, 31, 72, 98–105, 110–13, 115–18, 121–25, 127–29, 132, 140, 155–57, 161, 174 n. 7
emulsions, 16, 167 n. 10
endothermic reaction, 120–21
energy of motion, 114–15
energy resolution, 102–5, 108–10, 125–26
erg, 170, n. 42
Europa, ix
excitation, 115–16, 119
EXOSAT observatory, 80, 177 n. 8, 181 n. 17
exothermic reaction, 120–21
Exploration of the Solar System (NASA theme), 150–51
exposure time, 21, 25, 30, 62–63

Field, George, 146–47
Field report, 186 n. 5
film: chemical processing, xiii, 15–17, 58, 67
flares, 43, 72–75, 84–85
focal length, 157–59
foil mirrors, 158, 173 n.38, 189 n.12
formation flying, 158–59, 163, 189 n.14
frequency, 12–15, 39, 41–43, 94–97, 100–101, 118–19, 180 n. 7, 181 n. 23

galactic coordinates, 51–52, 172 n. 24
galaxies: *See* Ginga; Hickson Compact Group; Large Magellanic Cloud; M87; Markarian 421; Markarian 501; MCG-6-30-15; Milky Way; NGC 833; NGC 1553; NGC 3031 (M81); NGC 3516; NGC 4258; NGC 4697; NGC 4945; NGC 5128; NGC 5878; NGC 6814;

galaxies *(continued)*
PKS0637-75; Seyfert galaxies; Virgo cluster
galaxy, 9–10, 12, 25–34, 47–52, 54–55, 59–60, 65–66, 80–81, 85–86, 90–91, 93, 110–11, 120–22, 131–32, 137–38, 154, 157–61, 163, 168 n. 29, 178 n. 26; elliptical, 30–31, 47–49, 89–90, 159–60; halos, 29–30; infrared, 157; mass, 28, 33–34, 160–61; metallicity, 160; nucleus, 30–33; spiral, 29–30, 159–61. *See also* active galaxies; cluster (galaxies)
Galilean satellites (Io, Europa, Ganymede, and Callisto), ix
Galileo, 15–16
Gamma-ray, xiv, 12, 15–16, 23–24, 39–43, 62–63, 74, 121, 133–35, 154, 176 n.27
Gamma-ray Bursters (GRBs), 23–24
Ganymede, ix
gas, xiv, 3–4, 7, 10, 15–19, 21, 25–28, 29–31, 37–38, 47–52, 54–57, 59–62, 66–67, 70, 72–73, 80–81, 90, 92–100, 104–5, 110–12, 118–21, 124–27, 129–36, 138–40, 159–62, 242, 180 n. 9, 181 n. 13; accretion disk, 129–32; atoms, 84–85, 116–17, 142, 152–53; cloud, 117–18, 152–53, 180 n. 8, 181 n. 22; comet, 84–85, 175 n. 16; frozen, 82, 175 n. 16; hot, 7, 30–31, 34–35, 58–59, 61–62, 96–97, 127, 143–44, 148, 150–52, 154–56, 161–62, 186 n. 18; stellar, 70–71, 77–78, 87–88, 116, 129–30, 143–44, 148, 180 n. 9, 185 n. 8; ideal gas law, 177 n. 13
Gemini telescopes, 154–55
Generation-X (satellite), 159
GeV range, 41–43
gigahertz (GHz), 40
Ginga (galaxy), 31
glitch, 134–35
Goddard Space Flight Center (GSFC), x, 107–8, 173 n. 38
Goldin, Dan, 150
Goodrich Corporation, x–xi
Grande Jatte (G. Seurat), 37–38, 129–30
gravity, 10, 81–83, 88–89, 99–100, 105, 107–10, 114, 120–21, 131–32, 138–39, 150–51, 177 n. 15, 17, 178 n. 22, 186 n. 20, 189 n. 14
gravity well, 10, 105, 138–39

grazing incidence, 43–44, 149–51, 158
GRBs. *See* Gamma-ray Bursters
Great Observatory Program (NASA's), xiii–xiv, 41–43, 149–50. For Chandra, *see* Chandra observatory; for Compton GRO, *see* Compton (Gamma-Ray) observatory; for Hubble, *see* Hubble Space Telescope; for SIRTF, *see* Space Infrared Telescope Facility
GRO J1655-40 (X-ray nova), 85–86, 178 n. 27
GRO J1744-28 (X-ray binary), 90
ground state, 115
GRS1915+105 (X-ray binary), 87–88
Gursky, H., 21–22, 25, 167 n. 22
GX 339-4 (X-ray binary), 88, 179 n. 34

Hakucho (Swan), 31
halides, 16
Harvard College Observatory, 116–17, 167 n. 12
Harvard University, 105, 146–47, 184 n. 15
Hayashi, C., 60–61, 174 n. 1
HEAO (High Energy Astronomy Observatory), 30–31
HEAO-1, 30
Henize, K., 166 n. 11
Hertz, Heinrich, 170 n. 6
hertz, 39, 41
Hertzsprung, E., 116–17, 120
Hertzsprung-Russell diagram (H-R diagram), 116–17, 120, 126–27, 184 n. 15
Hester, J., 90–91
Hickson Compact Group 16 (galaxy cluster), 9–10, 66–67
High Energy Astronomy Observatory (HEAO), 30
high throughput, 154
High Throughput X-ray Satellite (Constellation-X), 154–56, 159, 188 n. 4, 189 n. 15
hot phase, 86–87, 95–98, 159–61
hot target, 23–24
H-R diagram. *See* Hertzsprung-Russell diagram
HST. *See* Hubble Space Telescope
HTXS. *See* High Throughput X-ray Satellite
Hubble, Edwin, 165 n. 4, 174 n. 7
Hubble Space Telescope, xiii–xiv, 3, 12–14, 44–45, 54, 65, 90–92, 137–38, 143, 149,

154–55, 162, 171 n. 16, 177 n. 6, 187 n. 10, 187 n. 13

Hulse, R., 186 n. 20

Hunter's Belt, 59–60

Huygens, C., 182 n. 24

Hyakutake (Comet), 70–72, 175 n. 18, 176 n. 21

IBM, 144–45

INTEGRAL (satellite), 43

intensity, 17, 64–65, 72–73, 94–95, 147–48, 169 n. 40, 175 n. 12, 180 n. 7, 185 n. 3

interferometer, 176 n. 20, 189 n. 14

interferometry, 162

intermediate polars, 156–57

International Space Station, 158–59

interstellar medium, 15–16, 30–31, 52, 73–74, 120, 131–33, 138, 159–61, 185 n. 7

inverse Compton scattering, 130, 176 n. 25

inverse square law, 64–65

Io, ix

ion, 17–18, 115–16, 119–20, 126, 130, 132, 185 n. 7

ionization, xiv, 17–18, 72, 86–87, 115–16, 119–20, 126, 132

Iridium constellation (satellites), 154–55, 188 n. 7

Israel, W., 182 n. 34

Jansky, Karl, 170 n. 7

jets, 21, 32–34, 43, 90–91, 110–11, 134–37, 157, 170 n. 44, 184 n. 13, 185 n. 4, 187 n. 19

June 1962 rocket launch, 19–21, 49–52, 112–13, 163, 167 n. 19

Jupiter, ix, 71–75, 82, 187 n. 19; magnetic field, 73

Keck telescopes, 154–55, 188 n. 3

Kellogg, E., 25–26

Kepler, J., 70

kilovolt, 10, 41, 61–62

Kirshner, R., 184 n. 15

Laplace, Pierre Simon, 59–60

Large Magellanic Cloud, 9–10, 31, 121–22, 133–34, 166 n. 12, 176 n. 27; (30 Doradus, 30 Doradus C), 9–10

Large Space Telescope, 149

Lawrence Livermore National Laboratory, 88–89

Laser Interferometer Space Antenna (LISA), 189 n. 14

Leonid Meteor, 54–55

light travel time argument, 21, 34, 78–79

lightning, 3–4, 25–26

light-year, 170 n. 48

LINEAR (1999/S4) (comet), 72–73

LINEAR (Lincoln Near Earth Asteroid Research project), 175 n. 19, 22

Lockheed-Martin, 21

Lockman, F., 172 n. 27

Log N–log S Relation, 54–55

look-back time, 161

Los Alamos National Laboratory, 23

LP 944-20 (brown dwarf), 74–75

luminosity, 4–5, 31–32, 48–51, 61–62, 64–65, 100, 127, 170 n. 41–43, 174 n. 9, 183 n. 11, 15, 185 n. 8

Luxe Calme et Volupté (H. Matisse), 37

M87 (active galaxy), 111

MACHOs (Massive Compact Halo Objects), 29–30, 169 n. 37

magnetic fields, 61, 71–75, 88–94, 133–37, 156–57, 179 n. 39, 185 n. 10, 12

magneto-hydrodynamics, 136–37

Markarian 421 (blazar), 43

Markarian 501 (blazar), 43

Mars, ix, 70, 149–50, 187 n. 19; exploration satellites (Mars Climate Observer, Mars Climate Orbiter, Mars Global Surveyor, Mars Observer, Mars Odyssey, Mars Pathfinder, Mars Polar Lander), 150, 187 n. 14, 16

Marshall Space Flight Center, x–xi, 67, 174 n. 11

Massachusetts Institute of Technology, x–xi, 88, 90–91, 174 n. 11, 176 n. 22, 179 n. 34

Massive Compact Halo Objects (MACHO), 29–30, 169 n. 37

Matisse, Henri, 37

matter flow, 61, 83, 86–87, 90–91, 160

Max Planck Institute, x–xi

MAXIM (Micro-Arc-second X-ray Imaging Mission), 162–63

Maxwell, James Clerk, xiii, 12–15
McClintock, J., 84–85, 178 n. 25
MCG-6-30-15 (Seyfert galaxy), 104–5
megahertz (MHz), 40
metallicity, 160
Metzger, A., 73
Micro-Arc-second X-ray Imaging Mission (MAXIM), 162–63
microns, 40
micro-quasars, 110–11
Milky Way, 29–30, 111, 170 n. 7, 172 n. 24
Miller, J., 110–11, 182 n. 30
mirror, 4–7, 32–33, 38–39, 43–47, 49–51, 62, 67–70, 125–26, 129–30, 147–52, 154–56, 158–59; 171 n. 16, 172 n. 17–18, 175 n. 12, 180 n. 10, 186 n.7, 187 n.10, 188 n. 24, 3; curved, 43–44; foil, 158, 173 n. 38, 189 n. 12; parabolic, 39–40; quality, 39, 45–47, 129–30; shells, 151
mission creep, 148–50
MMT (Multiple Mirror Telescope), 154–55, 188 n. 3
molecular clouds, 59–61, 138
moon, x, 15–16, 19–22, 49–52, 82–83, 162, 167 n. 21, 170 n. 48, 178 n.18, 180 n. 10, 187 n.12
Muller, Alex, 144–45

N157B (supernova remnant), 9–10, 166 n. 11
NASA, x–xi, 5–7, 30, 67, 88–89, 107–8, 129, 143, 145–46, 148–53, 159, 165 n. 3, 173 n. 31–32, 173 n. 38, 174 n. 11, 177 n. 5, 178 n. 21, 179 n. 37, 184 n. 18, 187 n. 16–17, 188 n. 20, 188 n. 22–23, 4–6, 189 n.14
National Radio Astronomy Observatory, 162, 172 n. 27
National Research Council of the National Academy of Sciences, 146–47
Naval Research Laboratory, 19
Near Earth Asteroid Tracking (NEAT), 175 n. 19, 176 n. 22
nebula, 21–22, 59–60, 65–66, 90–92, 118, 167 n. 11; catalog, 172 n. 20. *See also* Crab Nebula; Eagle Nebula; Orion Nebula
Neptune, x

neutrino, 28–30, 168 n.32, 169 n. 33–36, 177 n. 9, 184 n. 13, 186 n. 20
neutron, 28–29, 118–19, 126, 182 n. 5, 183 n. 6
neutron star, 7, 65, 74–85, 87–90, 111–12, 132–35, 176 n. 24, 28, 177 n. 13–14, 178 n. 22, 182 n. 5, 185 n. 8, 186 n.17, 189 n. 14
Newton, Isaac, 9–10, 115
Newton observatory, x, xiii–xiv, 3, 8–10, 39–40, 54, 66–67, 88–90, 102–4, 106–8, 112–13, 115, 124–25, 127–28, 143, 152–54, 156–61, 165 n. 1, 173 n. 28, 188 n. 24
NGC 833 (AGN), 10
NGC 1553 (elliptical galaxy), 48–49, 172 n. 22
NGC 3031 (M81), 184 n. 14
NGC 3516 (Seyfert galaxy), 104–5, 181 n. 19, 21
NGC 4258 (Seyfert galaxy), 111–12, 182 n. 34
NGC 4697 (elliptical galaxy), 48–49
NGC 4945 (Seyfert galaxy), 111
NGC 5128 (radio galaxy), 135–36
NGC 5878, 181 n. 15
NGC 6814 (Seyfert galaxy), 33–36
non-thermal processes, 130, 133
novae, 78–79, 84–86, 156–57, 166 n. 9, 178 n. 28; dwarf, 85–87, 156–57; X-ray, 85–86
nuclear: chemistry, 182 n. 5; energy, 65–66; reactions, 28–29, 60–61, 65–66, 81, 99–100, 165 n. 1, 168 n. 32, 176 n. 24; region, 157; weapons, 23–24
nucleo-chemical evolution, 140

O star, 87–88
observation: infrared, 31–32; optical, 21–22, 31–32; radio, 21–22, 31–33; ultra-violet, 31–32, 171 n. 16, X-ray, 179 n. 34
observatory, x, xiii–xiv, 23–24, 30, 54–55, 79–80, 88–90, 149–50, 156–57, 167 n. 12, 186 n. 6, 186 n. 20; gravity wave, 186 n. 20, 189 n. 14; neutrino, 169 n. 34, 186 n. 20
Office of Space Science, x–xi, 88–89, 150–51, 179 n. 37, 187 n. 16
Onnes, H.K., 144
optical band, 22–23, 26, 39–40, 61–62, 80–82, 126–27, 138, 140, 174 n. 3, 175 n. 19

optical catalog, 22–23, 30–31
Optical Coating Laboratory, Inc, x–xi
optical photon, 39–40, 129, 174 n. 3, 181 n. 21, 184 n. 1
optical spectra, 100–101, 121, 129–30, 132, 180 n. 10, 181 n. 15
orbit, 4–5, 34–36, 41–43, 54–55, 69–70, 78–80, 82–88, 109–10, 112–15, 132–33, 147–49, 175 n. 15, 177 n. 5–6, 186 n. 20; aphelion, 70; apogee, 70, 177 n. 8; atomic, 116–19; geosynchronous, 156; perigee, 70, 177 n. 8; perihelion, 70; shapes, 69–70
orbital period, 36, 70, 78–80, 177 n. 8
Orbiting Solar Observatory (OSO), 23, 168 n. 24; (OSO-3, OSO-5, OSO-7, OSO-8), 168 n. 24
orbits of: Astro-E, 10–11, 166 n. 14; Chandra, 5–7; Columbia, 4–5; comet, 70; Earth, 34, 70, 179 n. 35; Einstein, 30–31; Galaxy, 31; Hubble, 171 n. 16, 177 n. 5; lunar, 52, 82; Mercury, 144; Pluto, 174 n. 2; ROSAT, 22–23; satellites, 23–24, 79–80, 149–50, 155–59, 166 n. 5, 177 n. 8, 187 n. 15, 188 n. 7; XMM, 8–9
Origins and Evolution of the Universe (NASA theme), 150–51
Orion Nebula, 59–60, 65–66; Sword, 59–60, Trapezium, 59–60
oscillation, 29–30
outburst, 78–79, 84–90

Palomar Observatories, 21–22, 174 n. 7
parabolic mirror, 39–40
Parkes Radio Telescope, 32–33
Parkes Survey (PKS), 32–33, 135–36
particle accelerator, 18–19, 135–36, 167 n. 17, 185 n. 9
particle beam, 18–19
particle physics, 28–29
Pauli, Wolfgang, 28–29
Pegasus (Tenma) satellite, 31
perigee, 70, 177 n. 8
perihelion, 70
periodic table, 74, 118–21, 125–26, 137–38, 183 n. 7
Perseus (galaxy cluster), 186 n. 19
photographic emulsion, 167 n. 10

photographic film, 16, 67
photographic plates, 15–17, 58, 67
photoionization, 132
photon, 12, 15–16, 37, 39–40, 43–47, 57, 61–64, 67–68, 70–71, 76–80, 116, 119, 122–24, 129–30, 132, 134–35, 147–50, 154, 160–61, 168 n. 32, 174 n. 3, 175 n. 12, 176 n. 25, 181 n. 20–21, 184 n. 1
physicist, 12–14, 18–21, 28–29, 81, 83, 105, 114–18, 125–26, 138–39, 144, 167 n. 17, 180 n. 5, 182 n. 3
pixel, 148, 173 n. 37, 181 n. 18
PKS (Parkes Survey) 32–33, 135–36
PKS0637-75 (quasar), 32–33
Planck, Max, 180 n. 7
Planck curve, 95–97, 180 n. 7, 185 n. 2
Planck spectrum, 95–99, 171 n. 10, 180 n. 7, 185 n. 2
Planck's constant, 171 n. 10, 180 n. 7
planetary astronomy, ix
Pluto, x, 174 n. 2
point source, 7, 9–10, 24–25, 32–33, 45–46, 161, 178 n. 18
point spread function, 5–7, 36–39, 45–47, 58, 61, 66, 122–24, 129, 151–53, 158, 160–62, 242, 175 n. 12
pointillism, 37–39
polarization, 136–37, 166 n. 1, 186 n. 16
polarized light, 136–37
polars (polarized stars), 136–37; intermediate, 156–57
pollution, 160
Popper, D., 105
primordial matter, 25–26
Princeton University, 28–29, 137–38, 146–47, 185 n. 13, 187 n. 13
proportional counter, 15, 17–19, 31, 34–35, 45–47, 49–51, 88, 100–102, 151–52, 167 n. 16
proton, 28–29, 72, 118–19, 125–26, 168 n. 32, 182 n. 5, 183 n. 6
proton-proton chain, 168 n. 32
proto-star, 60–62, 66–67, 74–75, 159–60
pulsar, 74, 81–82, 90–91, 134–35, 176 n. 28; 1913+16, 186 n. 20

quantum, 114–16, 119–20; theory of gravity, 178 n. 22
quantum physics, 137–38, 181 n. 14, 182 n. 3

quasars, 65–67, 144, 157; micro, 110–11, 174 n. 7
quasi-radial accretion, 132–33

radial accretion, 132–33
radiation, 32–33, 41–43, 59–62, 74, 81, 86–87, 95–100, 109–10, 117–21, 126–27, 129–30, 132–33, 138–39, 144–45, 157, 177 n. 17, 179 n. 35, 180 n. 7, 185 n. 9
radio astronomer, 40, 142, 162, 172 n. 27
radio waves, 39–41, 43, 171 n. 15
Rapid Burster, 89–90
rarefaction, 130–31
Rayleigh-Taylor instability, 131–32
Raytheon Optical Systems, Inc, x–xi
red star, 84–86, 179 n. 30
relativistic jets, 135–36
remnant, supernova. *See* "remnant" under supernova
Request for Proposal, 146
resonance transition, 115–16
Roche: critical surface, 82–83, 177 n. 17; limit, 83, 178 n. 18; lobe, 82–83, 87–88, 132–33, 178 n. 19
rocket, 3–4, 8–11, 19–23, 39–40, 49–52, 76–77, 100–102, 112–13, 145, 147–48, 154–56, 158–59, 163, 166 n. 4, 167 n. 19, 181 n. 17, 187 n. 15
Roentgen, Wilhelm, 12–15, 17, 21, 49–51, 166 n. 2–4, 167 n. 11
Roentgensatellit satellite (ROSAT), 12–14, 22–23, 34–35, 45–47, 49–51, 55–56, 61, 70–73, 81–82, 90–91, 107–8, 112–13, 126–27, 152–53, 160, 166 n. 2
Rossi, Bruno, 179 n. 34
Rossi X-ray Timing Explorer (RXTE), 43, 88–90, 100–102, 111–13, 151–52, 157–58, 177 n. 6, 179 n. 35, 188 n. 23
rotation, 78–79, 109–11; accretion disk, 111–13; galaxy, 28; period, 77–79; pulsar, 90–91; rate, 28, 127–28; satellite, 25; stellar, 127–28
Russell, H.N., 116–17, 120
RXTE. *See* Rossi X-ray Timing Explorer

Sagan, Carl, ix–x, 175 n. 16
Sandage, Allan, 21–23
satellite, x, xiii–xiv, 8–9, 12–14, 18–19, 23–25, 31, 34–35, 39–46, 49–51, 70–71, 79–80, 88, 100–102, 133–34, 142–43, 145–59, 176 n. 20, 178 n. 26, 187 n. 17; 189 n. 14; aspect, 45–46; constellation, 149–50, 154–55; earth orbiting, 79–80, 156, 177 n. 5, 188 n. 7; free floating, 23–24; gamma-ray, 39–40; imaging, 49–51; large, 30, 154–55; life of, 145–46; mirrors, 148; observations, 19; proposals, 145–47; rotation, 25; scanning, 24–25; spinning, 23–25, 30, 188 n. 21; sun orbiting, 187 n. 15; weather, 177 n. 5, X-ray, xiv, 3, 8–11, 18–19, 23–24, 30–31, 34–35, 69–70, 76–77, 80, 85–86, 109–10, 143, 147–48, 153–56, 174 n. 8
satellites: *See* Astronomical Netherlands Satellite (ANS); Ariel V; Advanced Satellite for Cosmology and Astrophysics; Astro-E observatory; Orbiting Solar Observatory (OSO); BeppoSAX; Chandra observatory; High Throughput X-ray Satellite (Constellation-X); EXOSAT observatory; Ginga (Galaxy); Generation-X; HEAO (High Energy Astronomy Observatory); Iridium constellation; Newton observatory; Pegasus (Tenma) satellite; Roentgensatellit satellite (ROSAT); Rossi X-ray Timing Explorer (RXTE); Space Infrared Telescope Facility (SIRTF); Solar and Heliospheric Observatory (SOHO); Hakucho (Swan); Tracking and Data Relay Satellite; Uhuru satellite; Vela satellites; XEUS satellite; Yohkoh satellite
Saturn, x, 83
Schrieffer, J., 144
Schwarzschild, Martin, 33–34, 137–38
Schwarzschild Radius, 33–34
Sco X-1 (X-ray pulsar), 19–23, 100–102, 173 n. 29, 181 n. 17
Seliger, H., 14, 166 n. 4–6, 167 n. 11
Serlemitsos, Peter, 107–8, 189 n. 12
Seurat, Georges, 37–39, 129–30, 170 n. 2–3
Seyfert galaxies, 157
SGR (Soft Gamma-ray Repeaters), 133–35, 186 n. 14
Shapley, Harlow, 137–38
shock wave, 7, 74, 120, 130–32, 163, 185 n. 4

shuttle, 3–5, 112–13, 158–59, 173 n. 38, 177 n. 6

SIRTF. *See* Space Infrared Telescope Facility

silver (bromide, chloride, halides, iodide), 16

sky surveys, 17, 22–23, 25, 30–31, 49–51, 55–56, 61, 71–72, 81–82, 126–27

Skylab, 127–28, 133–34

Sloan Digital Sky Survey, 17, 167 n. 14

Smithsonian Astrophysical Observatory, x–xi, 84–85, 174 n. 11

SN1987A (supernova), 121–22, 176 n. 27, 184 n. 15

SN1988H (supernova), 181 n. 15

SN1993J (supernova), 184 n. 14

SNO (Sudbury Neutrino Observatory), 29, 169 n. 35

SNR. *See* "remnant" under supernova

soft gamma-ray repeaters (SGR), 133–35, 186 n. 14

Solar and Heliospheric Observatory (SOHO), 184 n. 18

solar luminosity, 31–32, 170 n. 42

solar models, 29

solar system, ix, 23–26, 32–33, 52, 60–61, 72–73, 116–17, 133–34, 137–38, 150–51, 173 n. 29, 174 n. 2, 176 n. 23, 187 n. 19

solar wind, 72–73, 179 n. 36

Southwest Research Institute, 73

Space Infrared Telescope Facility (SIRTF), xiii–xiv, 149–50, 187 n. 15

Space Research Organization Netherlands, x–xi

spatial resolution, 18–19, 34–35, 52, 73, 81–82, 90–91, 107, 121–24, 129–30, 139–40, 151–53, 157–58, 160–62, 167 n. 21, 171 n. 16, 173 n. 38

spectral resolution, 52, 58, 67, 71–72, 101–106, 108–10, 111, 121–26, 129, 151–53, 157–58, 173 n. 38

spectral types, 116–17, 126–28

spectroscopy, 4–5, 30–32, 74, 91–95, 98–99, 101–9, 112–16, 118, 120, 124–26, 137–38, 156–57, 177 n. 5, 180 n. 5

Spectrum X, 165 n. 1

speed, 117–18; of light, 32–35, 41, 76, 90–91, 105–6, 161, 170 n. 44–45, 174 n. 7, 180 n. 7–8, 181 n. 22–23; of sound, 131, 139–40, 185 n. 4

Spitzer, Lyman Jr., 187 n. 13

SS Cygni (SS Cyg) (cataclysmic variable), 86–87, 179 n. 32

star, 7, 23, 60–61, 74, 79–86, 93–101, 116, 120–22, 126–27, 132–33, 137–38, 171 n. 9, 178 n. 26, 180 n. 8, 183 n. 11, 184 n.15; aging, 74, 160; A star, 99–100; atmosphere, 127; binary, 23, 36, 81–85, 87–88, 102–4, 110, 112–13; blue, 21–22; brightness, 93, 116; B star, 87–88; Cas A, 7; chemistry, 118; clusters, 120; compact, 83–84, 132–33, 185 n. 8; constellation, 179 n. 32; core, 99–100, 120–21; donor, 83–84; evolution, 74, 81, 120–21, 131–32, 183 n. 10; exploded, 7, 9–10, 32–33, 52, 74, 81, 90, 176 n. 25, 177 n. 9, 184 n. 13; formation, 54–55, 59–61, 65–66, 92; fuel, 99–100, 120, 184 n. 15; hot, 24–25, 31–32; low-mass, 87–88, 121–22, 183 n. 11; mass, 7, 34, 81, 99–100, 160, 178 n. 23, 183 n. 11; naming schemes, 179 n. 32; near-star, 61; neutron, 7, 65, 74, 81–85, 87–90, 111–12, 132–34, 177 n. 13–14, 17, 178 n. 22, 185 n. 8; nova, 84–85; O star, 87–88; proto-star, 60–61, 74–75; quake, 134; receiving, 83–84; red, 84–86, 100; remnant, 9–10; shooting, 10–11; spectrum, 118; structure, 97–100, 120–21, 131–32, 144–45, 183 n. 11, 184 n. 12; supernova, 100–101; temperature, 119–20; trackers, 44–46, 148–49; VZ Sculptoris, 112–13; white dwarf, 99–100

Structure and Evolution of the Universe, 88–89, 150–51, 189 n. 14

Sudbury Neutrino Observatory (SNO), 29, 169 n. 35

Sunday Afternoon on the Island of La Grande Jatte (G. Seurat), 37

Sun-Earth Connection, 150–51

supernova, 7–10, 25–26, 31, 43, 52, 60–61, 73–74, 78–81, 90, 98–101, 120–21, 131–34, 138, 140, 159–60, 163, 166 n. 9, 176 n. 24, 178 n. 28, 181 n. 15, 189 n. 14; absorption (center), 100–101; creation, 121; emission, 140, 181 n. 21; remnant (SNR), 7, 9–10, 25–26, 43, 52,

supernova *(continued)*
    80–81, 90, 98–99, 120, 133–34, 140–42,
    159–60, 185 n. 5; spectrum, 100–101,
    121, 132; Type I, 121; Type II, 121; X-
    rays, 74, 131–32
supernovae. *See* Cassiopeia A; Large Ma-
    gellanic Cloud (30 Doradus C); N157B
    (remnant); SN1987A; SN1988H;
    SN1993J
Sword (Orion Nebula), 59–60

Taylor, J., 186 n. 20
telescope, xiii–xiv, 4–7, 15–16, 17–19, 30–
    31, 38–40, 57, 62–63, 68–70, 77–80,
    92–93, 106–7, 111–12, 138–39, 147–49,
    151–52, 154–58, 160–63, 171 n. 16, 175
    n. 12, 19, 187 n.13; amateur 186 n. 7,
    189 n. 17; collecting area, 23–24, 62–
    63, 65, 181 n. 20; cost, 151; diameter,
    158–59, 162; ground-based, 39–40,
    133, 171 n. 16; imaging, 36; infrared,
    43–44, 159–60; mirror, 4–5, 158; neu-
    trino, 186 n. 20; optical, xiii, 22–23,
    43–45, 143, 154–55, 173 n. 33, 188 n. 3;
    radio, 43–44, 170 n.7; ultraviolet, 43–
    44; X-ray, 5–7, 158–59, 173 n. 38, 184
    n. 18, 189 n. 13
Tenma (Pegasus), 31
TeV sources, 43
thermal: burst, 176 n. 25; motion, 168 n.
    30; processes, 130; radiation, 129–30,
    132, spectra, 180 n. 6
thermonuclear reactions, 74–75
torus, 110–11, 157
Tracking and Data Relay Satellite, 188 n. 22
transition, 114–16, 127–28, 130, 183 n. 8–
    9; downward, 114–15, 118–19, 185 n.
    7; electrons, 118–19; forbidden, 115–
    16; resonance, 115–16; upward, 114–
    15, 118–19
Transition Region and Coronal Explorer
    satellite (TRACE), 184 n. 18
Trapezium (Orion Nebula), 59–60
TRW Space and Electronic Group, x–xi,
    188 n. 5
Turner, Martin, 165 n. 1

Uhuru satellite, 23–25
universe: behavior, 28–29, 97–98, 161, 182
    n. 1; density, 138–40; development,
    25–26, 88–89, 137–40, 150–51, 161,
    174 n. 7, 176 n. 24; geometry, 55–56,
    138–39, 150–51; matter, 30
University of California (Berkeley), 29–30,
    169 n.37
University of Edinburgh, 29–30
University of Florida, 185 n. 12
University of Leicester, 165 n. 1
University of Michigan, 73
University of Texas, 184 n. 13
University of Wisconsin, 172 n. 25
upward transition, 114–15, 118–19
Uranus, x, 34–35
U.S. Congress, 39, 143, 146–49, 150–51,
    153, 167 n. 9, 187 n. 10, 12, 187 n. 14

V1432 Aquilae (V1432 Aql) (cataclysmic
    variable), 36
variability, 21, 25, 34, 54, 74–75, 79–81,
    147–48
Vela satellites, 23–24
velocity, 111, 116–18, 127, 174 n. 7, 185 n. 8
Very Large Array (VLA), 162
Vikings (Mars landers), ix, 150
Virgo galaxy cluster, 25, 111
voids, 25–26
voltage, 17–18, 108–9, 167 n. 8, 175 n. 14
VZ Sculptoris (VZ Scl) (cataclysmic vari-
    able), 112–13

Waite, J.H., 73
wavelength, 7, 15–16, 23–24, 26, 39–43,
    61–63, 66, 74–75, 94–102, 107–8, 112–
    13, 116, 119, 124–25, 130–31, 140,
    144–45, 147–50, 154, 162
Weakly Interacting Massive Particles
    (WIMPs), 28–29
weather: prediction, 145, 186 n. 3
Wexelblatt's scheduling algorithm, 187 n.
    11, 14
white dwarf, 29–30, 83–87, 99–100, 105,
    111–12, 137, 156–57, 178 n. 22, 181 n.
    22, 185 n. 8
WIMPs. *See* Weakly Interacting Massive
    Particles

XB1730-335 (Rapid Burster)(X-ray bi-
    nary), 89–90
XEUS (X-ray Evolving Universe Satellite),
    157–60, 163, 165 n. 1, 189 n. 11, 15

XMM (Newton), 8–10, 165 n. 1

X-ray astronomy, ix–x, xiii, 19–24, 25, 30–31, 56–57, 80, 112–13, 125–26, 128, 137–38, 142, 159–60, 165 n. 1, 167 n. 18, 189 n. 15

X-ray band, 23, 26–28, 32–34, 127, 139–40, 143, 160–61

X-ray binary, 31, 48–49, 87–91, 109–11, 159–61, 168 n. 29

X-Ray Calibration Facility (XRCF), 67–69, 175 n.12–14

X-ray calorimeter, 108–9, 151

X-ray catalog, 22–23

X-ray nova, 85–88

X-ray observatories. *See* Astro-E observatory; Chandra observatory; Einstein observatory; EXOSAT observatory; Newton observatory

X-ray observatory, x, 32–33, 136–37, 165 n. 1

X-ray Sky survey, 30, 71–72

XRCF (X-Ray Calibration Facility) 67–69, 175 n.12–14

XTE J1550-564 (X-ray nova), 88, 179 n. 34, 41

Yohkoh satellite, 184 n. 18

Zwicky, Fritz, 28